A PLUME BOOK

NUMBER

TOBIAS DANTZIG was born in Latvia in 1884. As a young man, he was caught distributing anti-Tsar propaganda and fled to Paris, where he studied under Henri Poincaré. He moved to the United States in 1910 and took a job as a lumberjack in the forests of Oregon. He received his Ph.D. in mathematics from Indiana University in 1916, and taught at Johns Hopkins, Columbia University, and the University of Maryland. He died in 1956.

JOSEPH MAZUR is professor of mathematics at Marlboro College, where he has taught a wide range of classes in all areas of mathematics, its history, and its philosophy. He is the author of *Euclid in the Rainforest*.

BARRY MAZUR is the Gerhard Gade University Professor at Harvard University. He teaches and does research in mathematics. He is the author of *Imagining Numbers (Particularly the Square Root of Minus Fifteen)*.

D0071566

TOBIAS DANTZIG

NUMBER

The Language of Science

Edited by
JOSEPH MAZUR

Foreword by
BARRY MAZUR

The MASTERPIECE SCIENCE *Edition*

A PLUME BOOK

PLUME
Published by Penguin Group
Penguin Group (USA) Inc., 375 Hudson Street, New York, New York 10014, USA • Penguin
Group (Canada), 90 Eglinton Avenue East, Suite 700, Toronto, Ontario, Canada M4P 2Y3 (a
division of Pearson Penguin Canada Inc.) • Penguin Books Ltd., 80 Strand, London WC2R
0RL, England • Penguin Ireland, 25 St. Stephen's Green, Dublin 2, Ireland (a division of
Penguin Books Ltd.) • Penguin Group (Australia), 250 Camberwell Road, Camberwell,
Victoria 3124, Australia (a division of Pearson Australia Group Pty. Ltd.) • Penguin Books
India Pvt. Ltd., 11 Community Centre, Panchsheel Park, New Delhi – 110 017, India •
Penguin Group (NZ), cnr Airborne and Rosedale Roads, Albany, Auckland 1310, New
Zealand (a division of Pearson New Zealand Ltd.) • Penguin Books (South Africa) (Pty.) Ltd.,
24 Sturdee Avenue, Rosebank, Johannesburg 2196, South Africa

Penguin Books Ltd., Registered Offices: 80 Strand, London WC2R 0RL, England

Published by Plume, a member of Penguin Group (USA) Inc. Previously published in a Pi Press
edition.

First Plume Printing, February 2007
10 9 8 7 6 5 4 3 2 1

This edition is a republication of the 4th edition of *Number,* originally published by Scribner,
an imprint of Simon & Schuster Inc.

Ⓟ REGISTERED TRADEMARK—MARCA REGISTRADA

CIP data is available.
ISBN 0-13-185627-8 (hc.)
ISBN 978-0-452-28811-9 (pbk.)

Printed in the United States of America

Contents

Contents

Foreword

The book you hold in your hands is a many-stranded meditation on Number, and is an ode to the beauties of mathematics.

This classic is about the evolution of the Number concept. Yes: Number has had, and will continue to have, an *evolution*. How did Number begin? We can only speculate.

Did Number make its initial entry into language as an adjective? Three cows, three days, three miles. Imagine the exhilaration you would feel if you were the first human to be struck with the startling thought that a unifying thread binds "three cows" to "three days," and that it may be worthwhile to deal with their common three-ness. This, if it ever occurred to a single person at a single time, would have been a monumental leap forward, for the disembodied concept of three-ness, the noun *three*, embraces far more than cows or days. It would also have set the stage for the comparison to be made between, say, one day and three days, thinking of the latter duration as triple the former, ushering in yet another view of *three*, in its role in the activity of tripling; *three* embodied, if you wish, in the verb *to triple*.

Or perhaps Number emerged from some other route: a form of incantation, for example, as in the children's rhyme "One, two, buckle my shoe...."

However it began, this story is still going on, and Number, humble Number, is showing itself ever more central to our understanding of what *is*. The early Pythagoreans must be dancing in their caves.

If I were someone who had a yen to learn about math, but never had the time to do so, and if I found myself marooned on that proverbial "desert island," the one book I would hope to have along is, to be honest, a good swimming manual. But the second book might very well be this one. For Dantzig accomplishes these essential tasks of scientific exposition: to assume his readers have no more than a general educated background; to give a clear and vivid account of material most essential to the story being told; to tell an important story; and—the task most rarely achieved of all—to explain ideas and not merely allude to them.

One of the beautiful strands in the story of Number is the manner in which the concept changed as mathematicians expanded the republic of numbers: from the counting numbers

$$1, 2, 3,\ldots$$

to the realm that includes negative numbers, and zero

$$\ldots -3, -2, -1, 0, +1, +2, +3, \ldots$$

and then to fractions, real numbers, complex numbers, and, via a different mode of colonization, to infinity and the hierarchy of infinities. Dantzig brings out the motivation for each of these augmentations: There is indeed a unity that ties these separate steps into a single narrative. In the midst of his discussion of the expansion of the number concept, Dantzig quotes Louis XIV. When asked what the guiding principle was of his international policy, Louis XIV answered, "Annexation! One can always find a clever lawyer to vindicate the act." But Dantzig himself does not relegate anything to legal counsel. He offers intimate glimpses of mathematical birth pangs, while constantly focusing on the vital question that hovers over this story: What does it mean for a mathematical object to exist? Dantzig, in his comment about the emergence of complex numbers muses that "For centuries [the concept of complex numbers] figured as a sort of mystic bond between reason and imagination." He quotes Leibniz to convey this turmoil of the intellect:

"[T]he Divine Spirit found a sublime outlet in that wonder of analysis, that portent of the ideal world, that amphibian between being and not-being, which we call the imaginary root of negative unity." (212)

Dantzig also tells us of his own early moments of perplexity:

"I recall my own emotions: I had just been initiated into the mysteries of the complex number. I remember my bewilderment: here were magnitudes patently impossible and yet susceptible of manipulations which lead to concrete results. It was a feeling of dissatisfaction, of restlessness, a desire to fill these illusory creatures, these empty symbols, with substance. Then I was taught to interpret these beings in a concrete geometrical way. There came then an immediate feeling of relief, as though I had solved an enigma, as though a ghost which had been causing me apprehension turned out to be no ghost at all, but a familiar part of my environment." (254)

The interplay between algebra and geometry is one of the grand themes of mathematics. The magic of high school analytic geometry that allows you to describe geometrically intriguing curves by simple algebraic formulas and tease out hidden properties of geometry by solving simple equations has flowered—in modern mathematics—into a powerful intermingling of algebraic and geometric intuitions, each fortifying the other. René Descartes proclaimed: "I would borrow the best of geometry and of algebra and correct all the faults of the one by the other." The contemporary mathematician Sir Michael Atiyah, in comparing the glories of geometric intuition with the extraordinary efficacy of algebraic methods, wrote recently:

"Algebra is the offer made by the devil to the mathematician. The devil says: I will give you this powerful machine, it will answer any question you like. All you need to do is give me your soul: give up geometry and you will have this marvelous machine. (Atiyah, Sir Michael. *Special Article: Mathematics in the 20th Century*. Page 7. Bulletin of the London Mathematical Society, 34 (2002) 1–15.)"

It takes Dantzig's delicacy to tell of the millennia-long courtship between arithmetic and geometry without smoothing out the Faustian edges of this love story.

In Euclid's *Elements of Geometry*, we encounter Euclid's definition of a line: "Definition 2. A line is breadthless length." Nowadays, we have other perspectives on that staple of plane geometry, the straight line. We have the number line, represented as a horizontal straight line extended infinitely in both directions on which all numbers—positive, negative, whole, fractional, or irrational—have their position. Also, to picture time variation, we call upon that crude model, the timeline, again represented as a horizontal straight line extended infinitely in both directions, to stand for the profound, ever-baffling, ever-moving frame of past/present/futures that we think we live in. The story of how these different conceptions of *straight line* negotiate with each other is yet another strand of Dantzig's tale.

Dantzig truly comes into his own in his discussion of the relationship between time and mathematics. He contrasts Cantor's theory, where infinite processes abound, a theory that he maintains is "frankly dynamic," with the theory of Dedekind, which he refers to as "static." Nowhere in Dedekind's definition of real number, says Dantzig, does Dedekind even "use the word *infinite* explicitly, or such words as *tend, grow, beyond measure, converge, limit, less than any assignable quantity*, or other substitutes."

At this point, reading Dantzig's account, we seem to have come to a resting place, for Dantzig writes:

"So it seems at first glance that here [in Dedekind's formulation of real numbers] we have finally achieved a complete emancipation of the number concept from the yoke of time." (182)

To be sure, this "complete emancipation" hardly holds up to Dantzig's second glance, and the eternal issues regarding time and its mathematical representation, regarding the continuum and its relationship to physical time, or to our lived time—problems we have been made aware of since Zeno—remain constant companions to the account of the evolution of number you will read in this book.

Dantzig asks: To what extent does the world, the scientific world, enter crucially as an influence on the mathematical world, and vice versa?

"The man of science will acts *as if* this world were an absolute whole controlled by laws independent of his own thoughts or act; but whenever he discovers a law of striking simplicity or one of sweeping universality or one which points to a perfect harmony in the cosmos, he will be wise to wonder what role his mind has played in the discovery, and whether the beautiful image he sees in the pool of eternity reveals the nature of this eternity, or is but a reflection of his own mind." (242)

Dantzig writes:

"The mathematician may be compared to a designer of garments, who is utterly oblivious of the creatures whom his garments may fit. To be sure, his art originated in the necessity for clothing such creatures, but this was long ago; to this day a shape will occasionally appear which will fit into the garment as if the garment had been made for it. Then there is no end of surprise and of delight!" (240)

This bears some resemblance in tone to the famous essay of the physicist Eugene Wigner, "The Unreasonable Effectiveness of Mathematics in the Natural Sciences," but Dantzig goes on, by offering us his highly personal notions of *subjective reality* and *objective reality*. Objective reality, according to Dantzig, is an impressively large receptacle including all the data that humanity has acquired (e.g., through the application of scientific instruments). He adopts Poincaré's definition of objective reality, "what is common to many thinking beings and could be common to all," to set the stage for his analysis of the relationship between Number and objective truth.

Now, in at least one of Immanuel Kant's reconfigurations of those two mighty words *subject* and *object,* a dominating role is played by Kant's delicate concept of the *sensus communis.* This *sensus communis* is an inner "general voice," somehow constructed within each of us, that gives us our expectations of how the rest of humanity will judge things.

The objective reality of Poincaré and Dantzig seems to require, similarly, a kind of inner voice, a faculty residing in us, telling us something about the rest of humanity: The Poincaré-Dantzig objective reality is a fundamentally subjective consensus of what is commonly held, or what *could be* held, to be objective. This view already alerts us to an underlying circularity lurking behind many discussions regarding objectivity and number, and, in particular behind the sentiments of the essay of Wigner. Dantzig treads around this lightly.

My brother Joe and I gave our father, Abe, a copy of *Number: The Language of Science* as a gift when he was in his early 70s. Abe had no mathematical education beyond high school, but retained an ardent love for the algebra he learned there. Once, when we were quite young, Abe imparted some of the marvels of algebra to us: "I'll tell you a secret," he began, in a conspiratorial voice. He proceeded to tell us how, by making use of the magic power of the cipher *X*, we could find *that number which when you double it and add one to it you get 11*. I was quite a literal-minded kid and really thought of *X* as our family's secret, until I was disabused of this attribution in some math class a few years later.

Our gift of Dantzig's book to Abe was an astounding hit. He worked through it, blackening the margins with notes, computations, exegeses; he read it over and over again. He engaged with numbers in the spirit of this book; he tested his own variants of the Goldbach Conjecture and called them his *Goldbach Variations*. He was, in a word, enraptured.

But none of this is surprising, for Dantzig's book captures both soul and intellect; it is one of the few great popular expository classics of mathematics truly accessible to everyone.

—*Barry Mazur*

Editor's Note to the Masterpiece Science Edition

The text of this edition of *Number* is based on the fourth edition, which was published in 1954. A new foreword, afterword, endnotes section, and annotated bibliography are included in this edition, and the original illustrations have been redrawn.

The fourth edition was divided into two parts. Part 1, "Evolution of the Number Concept," comprised the 12 chapters that make up the text of this edition. Part 2, "Problems Old and New," was more technical and dealt with specific concepts in depth. Both parts have been retained in this edition, only Part 2 is now set off from the text as appendixes, and the "part" label has been dropped from both sections.

In Part 2, Dantzig's writing became less descriptive and more symbolic, dealing less with ideas and more with methods, permitting him to present technical detail in a more concise form. Here, there seemed to be no need for endnotes or further commentaries. One might expect that a half-century of advancement in mathematics would force some changes to a section called "Problems Old and New," but the title is misleading; the problems of this section are not old or new, but are a collection of classic ideas chosen by Dantzig to show how mathematics is done.

In the previous editions of *Number*, sections were numbered within chapters. Because this numbering scheme served no function other than to indicate a break in thought from the previous paragraphs, the section numbers were deleted and replaced by a single line space.

Preface to the Fourth Edition

A quarter of the century ago, when this book was first written, I had grounds to regard the work as a pioneering effort, inasmuch as the *evolution of the number concept—* though a subject of lively discussion among professional mathematicians, logicians and philosophers—had not yet been presented to the general public as a cultural issue. Indeed, it was by no means certain at the time that there were enough lay readers interested in such issues to justify the publication of the book. The reception accorded to the work both here and abroad, and the numerous books on the same general theme which have followed in its wake have dispelled these doubts. The existence of a sizable body of readers who are concerned with the cultural aspects of mathematics and of the sciences which lean on mathematics is today a matter of record.

It is a stimulating experience for an author in the autumn of life to learn that the sustained demand for his first literary effort has warranted a new edition, and it was in this spirit that I approached the revision of the book. But as the work progressed, I became increasingly aware of the prodigious changes that have taken place since the last edition of the book appeared. The advances in technology, the spread of the statistical method, the advent of electronics, the emergence of nuclear physics, and, above all, the growing importance of automatic computors— have swelled beyond all expectation the ranks of people who live on the fringes of mathematical activity; and, at the same time, raised the general level of mathematical education. Thus was I confronted not only with a vastly increased audience, but with a far more sophisticated and exacting audience than the one I had addressed twenty odd years earlier. These sobering reflections had a decisive influence on the plan of this new edition. As to the extent I was able to meet the challenge of these changing times—it is for the reader to judge.

Except for a few passages which were brought up to date, the *Evolution of the Number Concept*, Part One of the present edition, is a verbatim reproduction of the original text. By contrast, Part Two—*Problems, Old and New*—is, for all intents and purposes, a new book. Furthermore, while Part One deals largely with concepts and ideas. Still, Part Two should not be construed as a commentary on the original text, but as an integrated story of *the development of method and argument in the field of number*. One could infer from this that the four chapters of *Problems, Old and New* are more technical in character than the original twelve, and such is indeed the case. On the other hand, quite a few topics of general interest were included among the subjects treated, and a reader skilled in the art of "skipping" could readily circumvent the more technical sections without straying off the main trail.

Tobias Dantzig

Pacific Palisades
California
September 1, 1953

Preface to the First Edition

This book deals with ideas, not with methods. All irrelevant technicalities have been studiously avoided, and to understand the issues involved no other mathematical equipment is required than that offered in the average high-school curriculum.

But though this book does not presuppose on the part of the reader a mathematical education, it presupposes something just as rare: a capacity for absorbing and appraising ideas.

Furthermore, while this book avoids the technical aspects of the subject, it is not written for those who are afflicted with an incurable horror of the symbol, nor for those who are inherently form-blind. This is a book on mathematics: it deals with symbol and form and with the ideas which are back of the symbol or of the form.

The author holds that our school curricula, by stripping mathematics of its cultural content and leaving a bare skeleton of technicalities, have repelled many a fine mind. It is the aim of this book to restore this cultural content and present the evolution of number as the profoundly human story which it is.

This is not a book on the history of the subject. Yet the historical method has been freely used to bring out the rôle intuition has played in the evolution of mathematical concepts. And so the story of number is here unfolded as a historical pageant of ideas, linked with the men who created these ideas and with the epochs which produced the men.

Can the fundamental issues of the science of number be presented without bringing in the whole intricate apparatus of the science? This book is the author's declaration of faith that it can be done. They who read shall judge!

Tobias Dantzig

Washington, D.C.
May 3, 1930

CHAPTER 1

Fingerprints

Ten cycles of the moon the Roman year comprised:
This number then was held in high esteem,
Because, perhaps, on fingers we are wont to count,
Or that a woman in twice five months brings forth,
Or else that numbers wax till ten they reach
And then from one begin their rhythm anew.
　　　　　　　　　　　　—Ovid, *Fasti, III.*

an, even in the lower stages of development, possesses a faculty which, for want of a better name, I shall call *Number Sense*. This faculty permits him to recognize that something has changed in a small collection when, without his direct knowledge, an object has been removed from or added to the collection.

Number sense should not be confused with counting, which is probably of a much later vintage, and involves, as we shall see, a rather intricate mental process. Counting, so far as we know, is an attribute exclusively human, whereas some brute species seem to possess a rudimentary number sense akin to our own. At least, such is the opinion of competent observers of animal behavior, and the theory is supported by a weighty mass of evidence.

Many birds, for instance, possess such a number sense. If a nest contains four eggs one can safely be taken, but when two are removed the bird generally deserts. In some unaccountable way the bird can distinguish two from three. But this faculty is by no

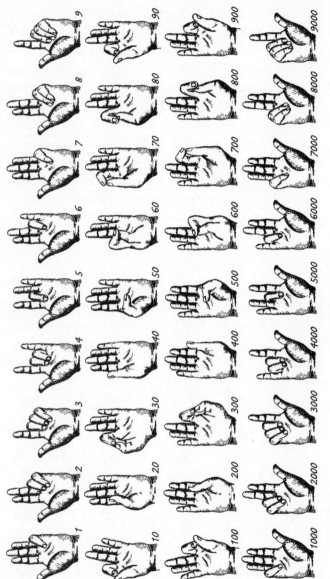

FINGER SYMBOLS
(FROM A MANUAL PUBLISHED IN 1520)

means confined to birds. In fact the most striking instance we know is that of the insect called the "solitary wasp." The mother wasp lays her eggs in individual cells and provides each egg with a number of live caterpillars on which the young feed when hatched. Now, the number of victims is remarkably constant for a given species of wasp: some species provide 5, others 12, others again as high as 24 caterpillars per cell. But most remarkable is the case of the *Genus Eumenus*, a variety in which the male is much smaller than the female. In some mysterious way the mother knows whether the egg will produce a male or a female grub and apportions the quantity of food accordingly; she does not change the species or size of the prey, but if the egg is male she supplies it with five victims, if female with ten.

The regularity in the action of the wasp and the fact that this action is connected with a fundamental function in the life of the insect make this last case less convincing than the one which follows. Here the action of the bird seems to border on the conscious:

A squire was determined to shoot a crow which made its nest in the watch-tower of his estate. Repeatedly he had tried to surprise the bird, but in vain: at the approach of the man the crow would leave its nest. From a distant tree it would watchfully wait until the man had left the tower and then return to its nest. One day the squire hit upon a ruse: two men entered the tower, one remained within, the other came out and went on. But the bird was not deceived: it kept away until the man within came out. The experiment was repeated on the succeeding days with two, three, then four men, yet without success. Finally, five men were sent: as before, all entered the tower, and one remained while the other four came out and went away. Here the crow lost count. Unable to distinguish between four and five it promptly returned to its nest.

Two arguments may be raised against such evidence. The first is that the species possessing such a number sense are exceedingly few, that no such faculty has been found among mammals, and that even the monkeys seem to lack it. The second argument is that in all known cases the number sense of animals is so limited in scope as to be ignored.

Now the first point is well taken. It is indeed a remarkable fact that the faculty of perceiving number, in one form or another, seems to be confined to some insects and birds and to men. Observation and experiments on dogs, horses and other domestic animals have failed to reveal any number sense.

As to the second argument, it is of little value, because the scope of the human number sense is also quite limited. In every practical case where civilized man is called upon to discern number, he is consciously or unconsciously aiding his direct number sense with such artifices as symmetric pattern reading, mental grouping or counting. *Counting* especially has become such an integral part of our mental equipment that psychological tests on our number perception are fraught with great difficulties. Nevertheless some progress has been made; carefully conducted experiments lead to the inevitable conclusion that the direct *visual* number sense of the average civilized man rarely extends beyond four, and that the *tactile* sense is still more limited in scope.

Anthropological studies on primitive peoples corroborate these results to a remarkable degree. They reveal that those savages *who have not reached the stage of finger counting* are almost completely deprived of all perception of number. Such is the case among numerous tribes in Australia, the South Sea Islands, South America, and Africa. Curr, who has made an extensive study of primitive Australia, holds that but few of the natives are able to discern four, and that no Australian in his wild state can perceive seven. The Bushmen of South Africa have no number words beyond *one*, *two* and *many*, and these words are so inarticulate that it may be doubted whether the natives attach a clear meaning to them.

We have no reasons to believe and many reasons to doubt that our own remote ancestors were better equipped, since practically all European languages bear traces of such early limitations. The English *thrice*, just like the Latin *ter*, has the double meaning: three times, and many. There is a plausible connection between the Latin *tres*, three, and *trans*, beyond; the same can be said regarding the French *très*, very, and *trois*, three.

The genesis of number is hidden behind the impenetrable veil of countless prehistoric ages. Has the concept been born of experience, or has experience merely served to render explicit what was already latent in the primitive mind: Here is a fascinating subject for metaphysical speculation, but for this very reason beyond the scope of this study.

If we are to judge of the development of our own remote ancestors by the mental state of contemporary tribes we cannot escape the conclusion that the beginnings were extremely modest. A rudimentary number sense, not greater in scope than that possessed by birds, was the nucleus from which the number concept grew. And there is little doubt that, left to this direct number perception, man would have advanced no further in the art of reckoning than the birds did. But through a series of remarkable circumstances man has learned to aid his exceedingly limited perception of number by an artifice which was destined to exert a tremendous influence on his future life. This artifice is counting, and it is to *counting* that we owe that extraordinary progress which we have made in expressing our universe in terms of number.

There are primitive languages which have words for every color of the rainbow but have no word for color; there are others which have all number words but no word for number. The same is true of other conceptions. The English language is very rich in native expressions for particular types of collections: *flock*, *herd*, *set*, *lot* and *bunch* apply to special cases; yet the words *collection* and *aggregate* are of foreign extraction.

The concrete preceded the abstract. "It must have required many ages to discover," says Bertrand Russell, "that a brace of pheasants and a couple of days were both instances of the number two." To this day we have quite a few ways of expressing the idea *two*: pair, couple, set, team, twin, brace, etc., etc.

A striking example of the extreme concreteness of the early number concept is the Thimshian language of a British Columbia tribe. There we find seven distinct sets of number words: one for flat objects and animals; one for round objects and time; one for counting men; one for long objects and trees; one for canoes; one for measures; one for counting when no definite object is referred to. The last is probably a later development; the others must be relics of the earliest days when the tribesmen had not yet learned to count.

It is counting that consolidated the concrete and therefore heterogeneous notion of plurality, so characteristic of primitive man, into the *homogeneous abstract number concept*, which made mathematics possible.

Yet, strange though it may seem, it is possible to arrive at a logical, clear-cut number concept without bringing in the artifices of counting.

We enter a hall. Before us are two collections: the seats of the auditorium, and the audience. *Without counting* we can ascertain whether the two collections are equal and, if not equal, which is the greater. For if every seat is taken and no man is standing, we *know without counting* that the two collections are equal. If every seat is taken and some in the audience are standing, *we know without counting* that there are more people than seats.

We derive this knowledge through a process which dominates all mathematics and which has received the name of *one-to-one correspondence*. It consists in assigning to every object of one collection an object of the other, the process being continued until one of the collections, or both, are exhausted.

The number technique of many primitive peoples is confined to just such such a matching or tallying. They keep the record of their herds and armies by means of notches cut in a tree or pebbles gathered in a pile. That our own ancestors were adept in such methods is evidenced by the etymology of the words *tally* and *calculate*, of which the first comes from the Latin *talea*, cutting, and the second from the Latin *calculus*, pebble.

It would seem at first that the process of correspondence gives only a means for comparing two collections, but is incapable of creating number in the absolute sense of the word. Yet the transition from relative number to absolute is not difficult. It is necessary only to create *model collections*, each typifying a possible collection. Estimating any given collection is then reduced to the selection among the available models of one which can be matched with the given collection member by member.

Primitive man finds such models in his immediate environment: the wings of a bird may symbolize the number two, clover-leaves three, the legs of an animal four, the fingers on his own hand five. Evidence of this origin of number words can be found in many a primitive language. Of course, once the *number word* has been created and adopted, it becomes as good a model as the object it originally represented. The necessity of discriminating between the name of the borrowed object and the number symbol itself would naturally tend to bring about a change in sound, until in the course of time the very connection between the two is lost to memory. As man learns to rely more and more on his language, the sounds supersede the images for which they stood, and the originally concrete models take the abstract form of number words. Memory and habit lend concreteness to these abstract forms, and so mere words become measures of plurality.

The concept I just described is called *cardinal* number. The cardinal number rests on the principle of correspondence: it implies *no counting*. To create a counting process it is not enough to have a motley array of models, comprehensive though this latter may be. We must devise a number *system*: our set of models must be arranged in an ordered sequence, a sequence which progresses in the sense of growing magnitude, the *natural sequence*: one, two, three…. Once this system is created, *counting a collection* means assigning to every member a term in the natural sequence in *ordered succession* until the collection is exhausted. The term of the natural sequence assigned to the *last* member of the collection is called the *ordinal number* of the collection.

The ordinal system may take the concrete form of a rosary, but this, of course, is not essential. The *ordinal* system acquires existence when the first few number words have been committed to memory in their *ordered succession,* and a phonetic scheme has been devised to pass from any larger number to its *successor.*

We have learned to pass with such facility from cardinal to ordinal number that the two aspects appear to us as one. To determine the plurality of a collection, i.e., its cardinal number, we do not bother any more to find a model collection with which we can match it,—we *count* it. And to the fact that we have learned to identify the two aspects of number is due our progress in mathematics. For whereas in practice we are really interested in the cardinal number, this latter is incapable of creating an arithmetic. The operations of arithmetic are based on the tacit assumption that *we can always pass from any number to its successor,* and this is the essence of the ordinal concept.

And so matching by itself is incapable of creating an art of reckoning. Without our ability to arrange things in ordered succession little progress could have been made. Correspondence and succession, the two principles which permeate all mathematics—nay, all realms of exact thought—are woven into the very fabric of our number system.

It is natural to inquire at this point whether this subtle distinction between cardinal and ordinal number had any part in the early history of the number concept. One is tempted to surmise that the cardinal number, based on matching only, preceded the ordinal number, which requires both matching and ordering. Yet the most careful investigations into primitive culture and philology fail to reveal any such precedence. Wherever any number technique exists at all, both aspects of number are found.

But, also, wherever a counting technique, worthy of the name, exists at all, *finger counting* has been found to either precede it or accompany it. And in his fingers man possesses a device which permits him to pass imperceptibly from cardinal to ordinal number. Should he want to indicate that a certain collection contains four objects he will raise or turn down four fingers *simultaneously;* should he want to count the same collection he will raise or turn down these fingers *in succession.* In the first case he is using his fingers as a cardinal model, in the second as an ordinal system. Unmistakable traces of this origin of count-ing are found in practically every primitive language. In most of these tongues the number "five" is expressed by "hand," the number "ten" by "two hands," or sometimes by "man." Further-more, in many primitive languages the number-words up to four are identical with the names given to the four fingers.

The more civilized languages underwent a process of attrition which obliterated the original meaning of the words. Yet here too "fingerprints" are not lacking. Compare the Sanskrit *pantcha*, five, with the related Persian *pentcha*, hand; the Russian "piat," five, with "piast," the outstretched hand.

It is to his *articulate ten fingers* that man owes his success in calculation. It is these fingers which have taught him to count and thus extend the scope of number indefinitely. Without this device the number technique of man could not have advanced far beyond the rudimentary number sense. And it is reasonable to conjecture that without our fingers the development of number, and consequently that of the exact sciences, to which we owe our material and intellectual progress, would have been hope-lessly dwarfed.

And yet, except that our children still learn to count on their fingers and that we ourselves sometimes resort to it as a gesture of emphasis, finger counting is a lost art among modern civilized people. The advent of writing, simplified numeration, and uni-versal schooling have rendered the art obsolete and superfluous. Under the circumstances it is only natural for us to underesti-mate the rôle that finger counting has played in the history of reckoning. Only a few hundred years ago finger counting was such a widespread custom in Western Europe that no manual of arithmetic was complete unless it gave full instructions in the method. (See page 2.)

The art of using his fingers in counting and in performing the simple operations of arithmetic, was then one of the accomplishments of an educated man. The greatest ingenuity was displayed in devising rules for adding and multiplying numbers on one's fingers. Thus, to this day, the peasant of central France (Auvergne) uses a curious method for multiplying numbers above 5. If he wishes to multiply 9×8, he bends down 4 fingers on his left hand (4 being the excess of 9 over 5), and 3 fingers on his right hand ($8 - 5 = 3$). Then the number of the bent-down fingers gives him the tens of the result ($4 + 3 = 7$), while the product of the unbent fingers gives him the units ($1 \times 2 = 2$).

Artifices of the same nature have been observed in widely separated places, such as Bessarabia, Serbia and Syria. Their striking similarity and the fact that these countries were all at one time parts of the great Roman Empire, lead one to suspect the Roman origin of these devices. Yet, it may be maintained with equal plausibility that these methods evolved independently, similar conditions bringing about similar results.

Even today the greater portion of humanity is counting on fingers: to primitive man, we must remember, this is the only means of performing the simple calculations of his daily life.

How old is our number language? It is impossible to indicate the exact period in which number words originated, yet there is unmistakable evidence that it preceded written history by many thousands of years. One fact we have mentioned already: all traces of the original meaning of the number words in European languages, with the possible exception of *five,* are lost. And this is the more remarkable, since, as a rule, number words possess an extraordinary stability. While time has wrought radical changes in all other aspects we find that the number vocabulary has been practically unaffected. In fact this stability is utilized by philologists to trace kinships between apparently remote language groups. The reader is invited to examine the table at the end of the chapter where the number words of the standard Indo-European languages are compared.

Why is it then that in spite of this stability no trace of the original meaning is found? A plausible conjecture is that while number words have remained unchanged since the days when they originated, the names of the concrete objects from which the number words were borrowed have undergone a complete metamorphosis.

As to the structure of the number language, philological researches disclose an almost universal uniformity. Everywhere the ten fingers of man have left their permanent imprint.

Indeed, there is no mistaking the influence of our ten fingers on the "selection" of the base of our number system. In all Indo-European languages, as well as Semitic, Mongolian, and most primitive languages, the base of numeration is ten, i.e., there are independent number words up to ten, beyond which some compounding principle is used until 100 is reached. All these languages have independent words for 100 and 1000, and some languages for even higher decimal units. There are apparent exceptions, such as the English *eleven* and *twelve*, or the German *elf* and *zwölf*, but these have been traced to *ein-lif* and *zwo-lif*, *lif* being old German for *ten*.

It is true that in addition to the decimal system, two other bases are reasonably widespread, but their character confirms to a remarkable degree the *anthropomorphic* nature of our counting scheme. These two other systems are the quinary, base 5, and the vigesimal, base 20.

In the *quinary* system there are independent number words up to *five,* and the compounding begins thereafter. (See table at the end of chapter.) It evidently originated among people who had the habit of counting on one hand. But why should man confine himself to one hand? A plausible explanation is that primitive man rarely goes about unarmed. If he wants to count, he tucks his weapon under his arm, the left arm as a rule, and counts on his left hand, using his right hand as check-off. This may explain why the left hand is almost universally used by right-handed people for counting.

Many languages still bear the traces of a quinary system, and it is reasonable to believe that some decimal systems passed through the quinary stage. Some philologists claim that even the Indo-European number languages are of a quinary origin. They point to the Greek word *pempazein,* to count by fives, and also to the unquestionably quinary character of the Roman numerals. However, there is no other evidence of this sort, and it is much more probable that our group of languages passed through a preliminary *vigesimal stage.*

This latter probably originated among the primitive tribes who counted on their toes as well as on their fingers. A most striking example of such a system is that used by the Maya Indians of Central America. Of the same general character was the system of the ancient Aztecs. The day of the Aztecs was divided into 20 hours; a division of the army contained 8000 soldiers ($8000 = 20 \times 20 \times 20$).

While pure vigesimal systems are rare, there are numerous languages where the decimal and the vigesimal systems have merged. We have the English *score, two-score,* and *three-score;* the *French vingt* (20) and *quatre-vingt* (4×20). The old French used this form still more frequently; a hospital in Paris originally built for 300 blind veterans bears the quaint name of

Quinze-Vingt (Fifteen-score); the name *Onze-Vingt* (Eleven-score) was given to a corps of police-sergeants comprising 220 men.

There exists among the most primitive tribes of Australia and Africa a system of numeration which has neither 5, 10, nor 20 for base. It is a *binary* system, i.e., of base two. These savages have not yet reached finger counting. They have independent numbers for one and two, and composite numbers up to six. Beyond six everything is denoted by "heap."

Curr, whom we have already quoted in connection with the Australian tribes, claims that most of these count by pairs. So strong, indeed, is this habit of the native that he will rarely notice that two pins have been removed from a row of seven; he will, however, become immediately aware if one pin is missing. His sense of *parity* is stronger than his number sense.

Curiously enough, this most primitive of bases had an eminent advocate in relatively recent times in no less a person than Leibnitz. A binary numeration requires but two symbols, 0 and 1, by means of which all other numbers are expressed, as shown in the following table:

Decimal	1	2	3	4	5	6	7	8
Binary	1	10	11	100	101	110	111	1000

Decimal	9	10	11	12	13	14	15	16
Binary	1001	1010	1011	1100	1101	1110	1111	10000

The advantages of the *base two* are economy of symbols and tremendous simplicity in operations. It must be remembered that every system requires that tables of addition and multiplication be committed to memory. For the binary system these reduce to $1 + 1 = 10$ and $1 \times 1 = 1$; whereas for the decimal, each table has 100 entries. Yet this advantage is more than offset by lack of compactness: thus the decimal number $4096 = 2^{12}$ would be expressed in the binary system by 1,000,000,000,000.

It is the mystic elegance of the binary system that made Leibnitz exclaim: *Omnibus ex nihil ducendis sufficit unum.* (One suffices to derive all out of nothing.) Says Laplace:

> "Leibnitz saw in his binary arithmetic the image of Creation ... He imagined that Unity represented God, and Zero the void; that the Supreme Being drew all beings from the void, just as unity and zero express all numbers in his system of numeration. This conception was so pleasing to Leibnitz that he communicated it to the Jesuit, Grimaldi, president of the Chinese tribunal for mathematics, in the hope that this emblem of creation would convert the Emperor of China, who was very fond of the sciences. I mention this merely to show how the prejudices of childhood may cloud the vision even of the greatest men!"

It is interesting to speculate what turn the history of culture would have taken if instead of flexible fingers man had had just two "inarticulate" stumps. If any system of numeration could at all have developed under such circumstances, it would have probably been of the binary type.

That mankind adopted the decimal system is a *physiological accident.* Those who see the hand of Providence in everything will have to admit that Providence is a poor mathematician. For outside its physiological merit the decimal base has little to commend itself. Almost any other base, with the possible exception of *nine,* would have done as well and probably better.

Indeed, if the choice of a base were left to a group of experts, we should probably witness a conflict between the practical man, who would insist on a base with the greatest number of divisors, such as *twelve,* and the mathematician, who would want a prime number, such as *seven* or *eleven,* for a base. As a matter of fact, late in the eighteenth century the great naturalist Buffon proposed that the duodecimal system (base 12) be universally adopted. He pointed to the fact that 12 has 4 divisors, while 10 has only two, and maintained that throughout the ages this inadequacy of our decimal system had been so keenly felt that, in spite of ten being the universal base, most measures had 12 secondary units.

On the other hand the great mathematician Lagrange claimed that a prime base is far more advantageous. He pointed to the fact that with a prime base every systematic fraction would be irreducible and would therefore represent the number in a unique way. In our present numeration, for instance, the decimal fraction .36 stands really for many fractions: 36/100, 18/50, and 9/25 .... Such an ambiguity would be considered lessened if a prime base, such as eleven, were adopted.

But whether the enlightened group to whom we would entrust the selection of the base decided on a prime or a composite base, we may rest assured that the number *ten* would not even be considered, for it is neither prime nor has it a sufficient number of divisors.

In our own age, when calculating devices have largely supplanted mental arithmetic, nobody would take either proposal seriously. The advantages gained are so slight, and the tradition of counting by tens so firm, that the challenge seems ridiculous.

From the standpoint of the history of culture a change of base, even if practicable, would be highly undesirable. As long as man counts by tens, his ten fingers will remind him of the human origin of this most important phase of his mental life. So may the decimal system stand as a living monument to the proposition:

Man is the measure of all things.

NUMBER WORDS OF SOME INDO-EUROPEAN LANGAUGES SHOWING THE
EXTRAORDINARY STABILITY OF NUMBER WORDS

	Sanskrit	Ancient Greek	Latin	German	English	French	Russian
1	eka	en	unus	eins	one	un	odyn
2	dva	duo	duo	zwei	two	deux	dva
3	tri	tri	tres	drei	three	trois	tri
4	catur	tetra	quatuor	vier	four	quatre	chetyre
5	panca	pente	quinque	fünf	five	cinq	piat
6	sas	hex	sex	sechs	six	six	shest
7	sapta	hepta	septem	sieben	seven	sept	sem
8	asta	octo	octo	acht	eight	huit	vosem
9	nava	ennea	novem	neun	nine	neuf	deviat
10	daca	deca	decem	zehn	ten	dix	desiat
100	cata	ecaton	centum	hundert	hundred	cent	sto
1000	sehastre	xilia	mille	tausend	thousand	mille	tysiaca

A TYPICAL QUINARY SYSTEM: THE
API LANGUAGE OF THE NEW
HEBRIDES

	Word	Meaning
1	tai	
2	lua	
3	tolu	
4	vari	
5	luna	hand
6	otai	other one
7	olua	" two
8	otolu	" three
9	ovair	" four
10	lua luna	two hands

A TYPICAL VIGESIMAL SYSTEM: THE
MAYA LANGUAGE OF CENTRAL
AMERICA

1	hun	1
20	kal	20
20^2	bak	400
20^3	pic	8000
20^4	calab	160,000
20^5	kinchel	3,200,000
20^6	alce	64,000,000

A TYPICAL BINARY SYSTEM: A WESTERN TRIBE OF TORRES STRAITS

1 urapun	3 okosa-urapun	5 okosa-okosa-urapun
2 okosa	4 okosa-okosa	6 okosa-okosa-okosa

CHAPTER 2

The Empty Column

*"It is India that gave us the ingenious method of express-
ing all numbers by means of ten symbols, each symbol
receiving a value of position as well as an absolute value;
a profound and important idea which appears so simple to
us now that we ignore its true merit. But its very simplicity
and the great ease which it has lent to all computations
put our arithmetic in the first rank of useful inventions;
and we shall appreciate the grandeur of this achievement
the more when we remember that it escaped the genius of
Archimedes and Apollonius, two of the greatest men produced by antiquity."*

—Laplace

As I am writing these lines there rings in my ears the old
refrain:

"Reading, 'Riting, 'Rithmetic,
Taught to the tune of a hickory-stick!"

In this chapter I propose to tell the story of one of three
R's, the one, which, though oldest, came hardest to mankind.

It is not a story of brilliant achievement, heroic deeds, or noble
sacrifice. It is a story of blind stumbling and chance discovery,
of groping in the dark and refusing to admit the light. It is a
story replete with obscurantism and prejudice, of sound judgment often eclipsed by loyalty to tradition, and of reason long
held subservient to custom. In short, it is a human story.

19

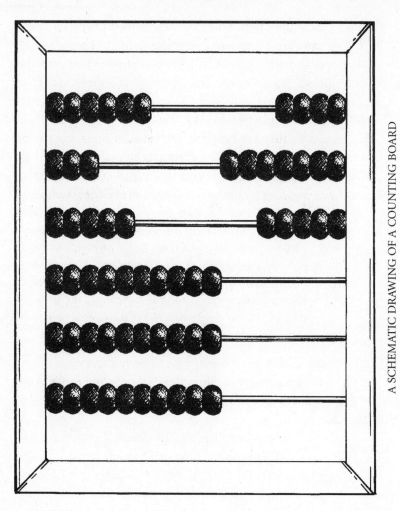

A SCHEMATIC DRAWING OF A COUNTING BOARD

Written numeration is probably as old as private property. There is little doubt that it originated in man's desire to keep a record of his flocks and other goods. Notches on a stick or tree, scratches on stones and rocks, marks in clay—these are the earliest forms of this endeavor to record numbers by written symbols. Archeological researches trace such records to times immemorial, as they are found in the caves of prehistoric man in Europe, Africa and Asia. Numeration is at least as old as written language, and there is evidence that it preceded it. Perhaps, even, the recording of numbers had suggested the recording of sounds.

The oldest records indicating the systematic use of written numerals are those of the ancient Sumerians and Egyptians. They are all traced back to about the same epoch, around 3500 B.C. When we examine them we are struck with the great similarity in the principles used. There is, of course, the possibility that there was communication between these peoples in spite of the distances that separated them. However, it is more likely that they developed their numerations along the lines of least resistance, i.e., that their numerations were but an outgrowth of the natural process of tallying. (See figure page 22.)

Indeed, whether it be the cuneiform numerals of the ancient Babylonians, the hieroglyphics of the Egyptian papyri, or the queer figures of the early Chinese records, we find everywhere a distinctly *cardinal* principle. Each numeral up to nine is merely a collection of strokes. The same principle is used beyond nine, units of a higher class, such as tens, hundreds, etc., being represented by special symbols.

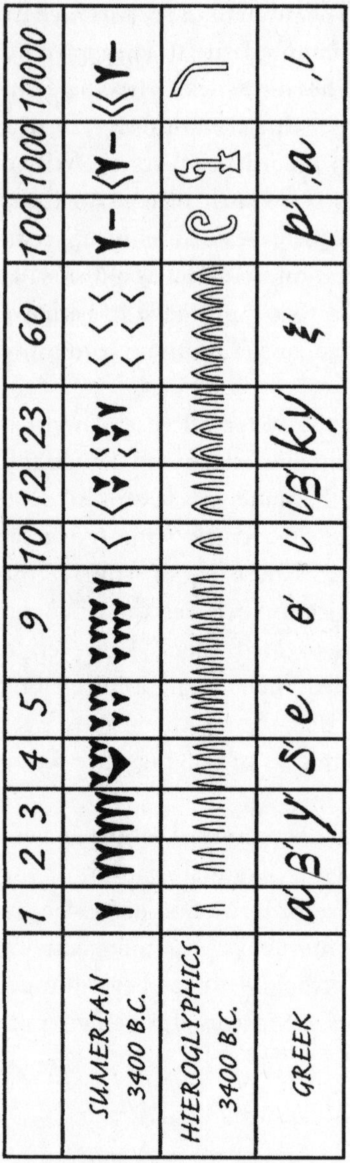

	1	2	3	4	5	9	10	12	23	60	100	1000	10000
SUMERIAN 3400 B.C.	𒁹	𒈫	𒐈	𒐉	𒐊	𒐗	⟨	⟨𒁹	⟨⟨𒐈	⟨⟨⟨	𒁹—	⟨𒁹—	⟨⟨𒁹—
HIEROGLYPHICS 3400 B.C.	∧	∧∧	∧∧∧	∧∧∧∧	∧∧∧∧∧	∧∧∧∧∧∧∧∧∧	⋀	⋀∧∧	⋀⋀∧∧∧	⋀⋀⋀⋀⋀⋀	℮	𓆼	⌐
GREEK	α΄	β΄	γ΄	δ΄	ε΄	θ΄	ι΄	ιβ΄	κγ΄	ξ	ρ΄	͵α	͵ι

ANCIENT NUMERATIONS

SCHEMATIC DRAWING OF ENGLISH TALLY-STICK

The English tally-stick of obscure but probably very ancient origin, also bears this unquestionably cardinal character. A schematic picture of the tally is shown in the accompanying figure. The small notches each represent a pound sterling, the larger ones 10 pounds, 100 pounds, etc.

It is curious that the English tally persisted for many centuries after the introduction of modern numeration made its use ridiculously obsolete. In fact it was responsible for an important episode in the history of Parliament. Charles Dickens described this episode with inimitable sarcasm in an address on Administrative Reform, which he delivered a few years after the incident occurred.

"Ages ago a savage mode of keeping accounts on notched sticks was introduced into the Court of Exchequer and the accounts were kept much as Robinson Crusoe kept his calendar on the desert island. A multitude of accountants, bookkeepers, and actuaries were born and died .... Still official routine inclined to those notched sticks as if they were pillars of the Constitution, and still the Exchequer accounts continued to be kept on certain splints of elm-wood called *tallies*. In the reign of George III an inquiry was made by some revolutionary spirit whether, pens, ink and paper, slates and pencils being in existence, this obstinate adherence to an obsolete custom ought to be continued, and whether a change ought not be effected. All the red tape in the country grew redder at the bare mention of this bold and original conception, and it took until 1826 to get these sticks abolished. In 1834 it was found that there was a considerable accumulation of them; and the question then arose, what was to be done with such worn-out, worm-eaten, rotten old bits of wood? The sticks were housed in Westminster, and it would naturally occur to any intelligent person that nothing could be easier than to allow them to be carried away for firewood by the miserable people who lived in that neighborhood. However, they never had been useful, and official routine required that they should never be, and so the order went out that they were to be privately and confidentially burned. It came to pass that they were burned in a stove in the House of Lords. The stove, over-gorged with these preposterous sticks, set fire to the panelling; the panelling set fire to the House of Commons; the two houses were reduced to ashes; architects were called in to build others; and we are now in the second million of the cost thereof."

As opposed to this purely cardinal character of the earliest records there is the ordinal numeration, in which the numbers are represented by the letters of an alphabet in their spoken succession.

The earliest evidence of this principle is that of the Phoenician numeration. It probably arose from the urge for compactness brought about by the complexities of a growing commerce. The Phoenician origin of both the Hebrew and the Greek numeration is unquestionable: the Phoenician system was adopted bodily, together with the alphabet, and even the sounds of the letters were retained.

On the other hand, the Roman numeration, which has survived to this day, shows a marked return to the earlier cardinal methods. Yet Greek influence is shown in the literal symbols adopted for certain units, such as X for ten, C for hundred, M for thousand. But the substitution of letters for the more picturesque symbols of the Chaldeans or the Egyptians does not constitute a departure from principle.

The evolution of the numerations of antiquity found its final expression in the ordinal system of the Greeks and the cardinal system of Rome. Which of the two was superior? The question would have significance if the only object of a numeration were a compact recording of quantity. But this is not the main issue. A far more important question is: how well is the system adapted to arithmetical operations, and what ease does it lend to calculations?

In this respect there is hardly any choice between the two methods: neither was capable of creating an arithmetic which could be used by a man of average intelligence. This is why, from the beginning of history until the advent of our modern *positional* numeration, so little progress was made in the art of reckoning.

Not that there were no attempts to devise rules for operating on these numerals. How difficult these rules were can be gleaned from the great awe in which all reckoning was held in these days. A man skilled in the art was regarded as endowed with almost supernatural powers. This may explain why arithmetic from time immemorial was so assiduously cultivated by the priesthood. We shall have occasion later to dwell at greater length on this relation of early mathematics to religious rites and mysteries. Not only was this true of the ancient Orient, where science was built around religion, but even the enlightened Greeks never completely freed themselves from this mysticism of number and form.

And to a certain extent this awe persists to this day. The average man identifies mathematical ability with quickness in figures. "So you are a mathematician? Why, then you have no trouble with your income-tax return!" What mathematician has not at least once in his career been so addressed? There is, perhaps, unconscious irony in these words, for are not most professional mathematicians spared all trouble incident to excessive income?

There is a story of a German merchant of the fifteenth century, which I have not succeeded in authenticating, but it is so characteristic of the situation then existing that I cannot resist the temptation of telling it. It appears that the merchant had a son whom he desired to give an advanced commercial education. He appealed to a prominent professor of a university for advice as to where he should send his son. The reply was that if the mathematical curriculum of the young man was to be confined to adding and subtracting, he perhaps could obtain the instruction in a German university; but the art of multiplying and dividing, he continued, had been greatly developed in Italy, which in his opinion was the only country where such advanced instruction could be obtained.

As a matter of fact, multiplication and division as practiced in those days had little in common with the modern operations bearing the same names. Multiplication, for instance, was a succession of *duplations,* which was the name given to the doubling of a number. In the same way division was reduced to *mediation,* i.e., "halving" a number. A clearer insight into the status of reckoning in the Middle Ages can be obtained from an example. Using modern notations:

Today	Thirteenth century
46	$46 \times 2 = 92$
13	$46 \times 4 = 92 \times 2 = 184$
——	
138	$46 \times 8 = 184 \times 2 = 368$
46	$368 + 184 + 46 = 598$
——	
598	

We begin to understand why humanity so obstinately clung to such devices as the abacus or even the tally. Computations which a child can now perform required then the services of a specialist, and what is now only a matter of a few minutes meant in the twelfth century days of elaborate work.

The greatly increased facility with which the average man today manipulates number has been often taken as proof of the growth of the human intellect. The truth of the matter is that the difficulties then experienced were inherent in the numeration in use, a numeration not susceptible to simple, clear-cut rules. The discovery of the modern positional numeration did away with these obstacles and made arithmetic accessible even to the dullest mind.

The growing complexities of life, industry and commerce, of landed property and slave-holding, of taxation and military organization—all called for calculations more or less intricate, but beyond the scope of the finger technique. The rigid, unwieldy numeration was incapable of meeting the demand. How did man, in the five thousand years of his civilized existence which preceded modern numeration, counter these difficulties?

The answer is that from the very outset he had to resort to mechanical devices which vary in form with place and age but are all the same in principle. The scheme can be typified by the curious method of counting an army which has been found in Madagascar. The soldiers are made to file through a narrow passage, and one pebble is dropped for each. When 10 pebbles are counted, a pebble is cast into another pile representing tens, and the counting continues. When 10 pebbles are amassed in the second pile, a pebble is cast into a third pile representing hundreds, and so on until all the soldiers have been accounted for.

From this there is but one step to the *counting board* or *abacus* which in one form or another has been found in practically every country where a counting technique exists. The abacus in its general form consists of a flat board divided into a series of parallel columns, each column representing a distinct decimal class, such as units, tens, hundreds, etc. The board is provided with a set of counters which are used to indicate the number of units in each class. For instance, to represent 574 on the abacus, 4 counters are put on the last column, 7 counters on the next to last and 5 on the third to the last column. (See figure, page 20.)

The many counting boards known differ merely in the construction of the columns and in the type of counters used. The Greek and Roman types had loose counters, while the Chinese Suan-Pan of today has perforated balls sliding on slender bamboo rods. The Russian Szczety, like the Chinese variety, consists of a wooden frame on which are mounted a series of wire rods with sliding buttons for counters. Finally, it is more than probable that the ancient Hindu *dust board* was also an abacus in principle, the part of the counters here being played by erasable marks written on sand.

The origin of the word abacus is not certain. Some trace it to the Semitic *abac,* dust; others believe that it came from the Greek *abax,* slab. The instrument was widely used in Greece, and we find references to it in Herodotus and Polybius. The latter, commenting on the court of Philip II of Macedonia in his *Historia* makes this suggestive statement:

> "Like counters on the abacus which at the pleasure of the calculator may at one moment be worth a talent and the next moment a chalcus, so are the courtiers at their King's nod at one moment at the height of prosperity and at the next objects of human pity."

To this day the counting board is in daily use in the rural districts of Russia and throughout China, where it persists in open competition with modern calculating devices. But in Western Europe and America the abacus survived as a mere curiosity which few people have seen except in pictures. Few realize how extensively the abacus was used in their own countries only a few hundred years ago, where after a fashion it managed to meet the difficulties which were beyond the power of a clumsy numeration.

One who reflects upon the history of reckoning up to the invention of the principle of position is struck by the paucity of achievement. This long period of nearly five thousand years saw the fall and rise of many a civilization, each leaving behind it a heritage of literature, art, philosophy and religion. But what was the net achievement in the field of reckoning, the earliest art practiced by man? An inflexible numeration so crude as to make progress well-nigh impossible, and a calculating device so limited in scope that even elementary calculations called for the services of an expert. And what is more, man used these devices for thousands of years without making a single worth-while improvement in the instrument, without contributing a single important idea to the system!

This criticism may sound severe; after all it is not fair to judge the achievements of a remote age by the standards of our own time of accelerated progress and feverish activity. Yet, even when compared with the slow growth of ideas during the Dark Ages, the history of reckoning presents a peculiar picture of desolate stagnation.

When viewed in this light, the achievement of the unknown Hindu who some time in the first centuries of our era discovered the *principle of position* assumes the proportions of a world-event. Not only did this principle constitute a radical departure in method, but we know now that without it no progress in arithmetic was possible. And yet the principle is so simple that today the dullest school boy has no difficulty in grasping it. In a measure, it is suggested by the very structure of our number language. Indeed, it would appear that the first attempt to translate the action of the counting board into the language of numerals ought to have resulted in the discovery of the principle of position.

Particularly puzzling to us is the fact that the great mathematicians of classical Greece did not stumble on it. Is it that the Greeks had such a marked contempt for applied science, leaving even the instruction of their children to the slaves? But if so, how is it that the nation which gave us geometry and carried this science so far, did not create even a rudimentary algebra? Is it not equally strange that algebra, that cornerstone of modern mathematics, also originated in India and at about the same time when positional numeration did?

A close examination of the anatomy of our modern numeration may shed light on these questions. The principle of position consists in giving the numeral a value which depends not only on the member of the natural sequence it represents, but also on the position it occupies with respect to the other symbols of the group. Thus, the same digit 2 has different meanings in the three numbers 342, 725, 269: in the first case it stands for two; in the second for twenty, in the third for two hundred. As a matter of fact 342 is just an abbreviation for three hundred plus four tens plus two units.

But that is precisely the scheme of the counting board, where 342 is represented by

And, as I said before, it would seem that it is sufficient to translate this scheme into the language of numerals to obtain substantially what we have today.

True! But there is one difficulty. Any attempt to make a permanent record of a counting-board operation would meet the obstacle that such an entry as ☰ = may represent any one of several numbers: 32, 302, 320, 3002, and 3020 among others. In order to avoid this ambiguity it is essential to have some method of representing the gaps, i.e., what is needed is a *symbol for an empty column.*

We see therefore that no progress was possible until a symbol was invented for an *empty* class, a symbol for *nothing,* our modern *zero.* The concrete mind of the ancient Greeks could not conceive the void as a number, let alone endow the void with a symbol.

And neither did the unknown Hindu see in zero the symbol of nothing. The Indian term for zero was *sunya,* which meant *empty* or *blank,* but had no connotation of "void" of "nothing." And so, from all appearances, the discovery of zero was an accident brought about by an attempt to make an unambiguous permanent record of a counting board operation.

How the Indian *sunya* became the zero of today constitutes one of the most interesting chapters in the history of culture. When the Arabs of the tenth century adopted the Indian numeration, they translated the Indian *sunya* by their own, *sifr,* which meant empty in Arabic. When the Indo-Arabic numeration was first introduced into Italy, *sifr* was latinized into *zephirum.* This happened at the beginning of the thirteenth century, and in the course of the next hundred years the word underwent a series of changes which culminated in the Italian *zero.*

About the same time Jordanus Nemararius was introducing the Arabic system into Germany. He kept the Arabic word, changing it slightly to *cifra.* That for some time in the learned circles of Europe the word *cifra* and its derivatives denoted zero is shown by the fact that the great Gauss, the last of the mathematicians of the nineteenth century who wrote in Latin, still used *cifra* in this sense. In the English language the word *cifra* has become *cipher* and has retained its original meaning of zero.

The attitude of the common people toward this new numeration is reflected in the fact that soon after its introduction into Europe, the word *cifra* was used as a secret sign; but this connotation was altogether lost in the succeeding centuries. The verb *decipher* remains as a monument of these early days.

The next stage in this development saw the new art of reckoning spread more widely. It is significant that the essential part played by zero in this new system did not escape the notice of the masses. Indeed, they identified the whole system with its most striking feature, the *cifra,* and this explains how this word in its different forms, *ziffer, chiffre,* etc., came to receive the meaning of numeral, which it has in Europe today.

This double meaning, the popular *cifra* standing for numeral and the *cifra* of the learned signifying zero, caused considerable confusion. In vain did scholars attempt to revive the original meaning of the word: the popular meaning had taken deep root.

The learned had to yield to popular usage, and the matter was eventually settled by adopting the Italian zero in the sense in which it is used today.

The same interest attaches to the word *algorithm*. As the term is used today, it applies to any mathematical procedure consisting of an indefinite number of steps, each step applying to the results of the one preceding it. But between the tenth and fifteenth centuries *algorithm* was synonymous with positional numeration. We now know that the word is merely a corruption of Al Kworesmi, the name of the Arabian mathematician of the ninth century whose book (in Latin translation) was the first work on this subject to reach Western Europe.

Today, when positional numeration has become a part of our daily life, it seems that the superiority of this method, the compactness of its notation, the ease and elegance it introduced in calculations, should have assured the rapid and sweeping acceptance of it. In reality, the transition, far from being immediate, extended over long centuries. The struggle between the *Abacists*, who defended the old traditions, and the *Algorists*, who advocated the reform, lasted from the eleventh to the fifteenth century and went through all the usual stages of obscurantism and reaction. In some places, Arabic numerals were banned from official documents; in others, the art was prohibited altogether. And, as usual, *prohibition* did not succeed in abolishing, but merely served to spread *bootlegging*, ample evidence of which is found in the thirteenth century archives of Italy, where, it appears, merchants were using the Arabic numerals as a sort of secret code.

Yet, for a while reaction succeeded in arresting the progress and in hampering the development of the new system. Indeed, little of essential value or of lasting influence was contributed to the art of reckoning in these transition centuries. Only the outward appearance of the numerals went through a series of changes; not, however, from any desire for improvement, but because the manuals of these days were hand-written. In fact, the numerals did not assume a stable form until the introduction of printing. It can be added parenthetically that so great was the stabilizing influence of printing that the numerals of today have essentially the same appearance as those of the fifteenth century.

As to the final victory of the Algorists, no definite date can be set. We do know that at the beginning of the sixteenth century the supremacy of the new numeration was incontestable. Since then progress was unhampered, so that in the course of the next hundred years all the rules of operations, both on integers and on common and decimal fractions, reached practically the same scope and form in which they are taught today in our schools.

Another century, and the Abacists and all they stood for were so completely forgotten that various peoples of Europe began each to regard the positional numeration as its own national achievement. So, for instance, early in the nineteenth century we find that Arabic numerals were called in Germany *Deutsche* with a view to differentiating them from the *Roman,* which were recognized as of foreign origin.

As to the abacus itself, no traces of it are found in Western Europe during the eighteenth century. Its reappearance early in the nineteenth century occurred under very curious circumstances. The mathematician Poncelet, a general under Napoleon, was captured in the Russian campaign and spent many years in Russia as a prisoner of war. Upon returning to France he brought among other curios, a Russian abacus. For many years to come, this importation of Poncelet's was regarded as a great curiosity of "barbaric" origin. Such examples of national amnesia abound in the history of culture. How many educated people even today know that only four hundred years ago finger counting was the average man's only means of calculating, while the counting board was accessible only to the professional calculators of the time?

Conceived in all probability as the symbol for an empty column on a counting board, the Indian *sunya* was destined to become the turning-point in a development without which the progress of modern science, industry, or commerce is inconceivable. And the influence of this great discovery was by no means confined to arithmetic. By paving the way to a generalized number concept, it played just as fundamental a rôle in practically every branch of mathematics. In the history of culture the discovery of zero will always stand out as one of the greatest single achievements of the human race.

A great discovery! Yes. But, like so many other early discoveries, which have profoundly affected the life of the race,—not the reward of painstaking research, but a gift from blind chance.

CHAPTER 3

Number-lore

"What is beautiful and definite and the object of knowledge is by nature prior to the indefinite and the incomprehensible and the ugly."

—Nicomachus

No two branches of mathematics present a greater contrast than arithmetic and the *Theory of Numbers*.

The great generality and simplicity of its rules makes arithmetic accessible to the dullest mind. In fact, facility in reckoning is merely a matter of memory, and the lightning calculators are but human machines, whose one advantage over the mechanical variety is greater portability.

On the other hand, the theory of numbers is by far the most difficult of all mathematical disciplines. It is true that the statement of its problems is so simple that even a child can understand what is at issue. But, the methods used are so individual that uncanny ingenuity and the greatest skill are required to find a proper avenue of approach. Here intuition is given free play. Most of the properties known have been discovered by a sort of *induction*. Statements held true for centuries have been later proved false, and to this day there are problems which have challenged the power of the greatest mathematicians and still remain unsolved.

Arithmetic is the foundation of all mathematics, pure or applied. It is the most useful of all sciences, and there is, probably, no other branch of human knowledge which is more widely spread among the masses.

On the other hand, the theory of numbers is the branch of mathematics which has found the least number of applications. Not only has it so far remained without influence on technical progress, but even in the domain of pure mathematics it has always occupied an isolated position, only loosely connected with the general body of the science.

Those who are inclined towards a utilitarian interpretation of the history of culture would be tempted to conclude that arithmetic preceded the theory of numbers. But the opposite is true. The theory of integers is one of the oldest branches of mathematics, while modern arithmetic is scarcely four hundred years old.

This is reflected in the history of the word. The Greek word *arithmos* meant number, and *arithmetica* was the theory of numbers even as late as the seventeenth century. What we call arithmetic today was *logistica* to the Greeks, and in the Middle Ages was called, as we saw, *algorism*.

But while the spectacular story which I am about to tell has little direct bearing on the development of other mathematical concepts, nothing could serve better to illustrate the evolution of these concepts.

The individual attributes of integers were the object of human speculation from the earliest days, while their more intrinsic properties were taken for granted. How do we account for this strange phenomenon?

The life of man, to borrow a famous maxim of Montesquieu, is but a succession of vain hopes and groundless fears. These hopes and fears which to this day find their expression in a vague and intangible religious mysticism, took in these early days much more concrete and tangible forms. Stars and stones, beasts and herbs, words and numbers, were symptoms and agents of human destiny.

The genesis of all science can be traced to the contemplation of these occult influences. Astrology preceded astronomy, chemistry grew out of alchemy, and the theory of numbers had its precursor in a sort of numerology which to this day persists in otherwise unaccountable omens and superstitions.

"For seven days seven priests with seven trumpets invested Jericho, and on the seventh day they encompassed the city seven times."

Forty days and forty nights lasted the rain which brought about the great deluge. For forty days and forty nights Moses conferred with Jehovah on Mount Sinai. Forty years were the children of Israel wandering in the wilderness.

Six, seven and forty were the ominous numbers of the Hebrews, and Christian theology inherited the seven: the seven deadly sins, the seven virtues, the seven spirits of God, seven joys of the Virgin Mary, seven devils cast out of Magdalen.

The Babylonians and Persians preferred sixty and its multiples. Xerxes punished the Hellespont with 300 lashes, and Darius ordered the Gyndes to be dug up into 360 ditches, because one of his holy horses had drowned in the river.

Religious values, says Poincaré, vary with longitude and latitude. While 3, 7, 10, 13, 40 and 60 were especially favored, we find practically every other number invested with occult significance in different places and at different times. Thus the Babylonians associated with each one of their gods a number up to 60, the number indicating the rank of the god in the heavenly hierarchy.

Strikingly similar to the Babylonian was the number worship of the Pythagoreans. It almost seems as if for fear of offending a number by ignoring it, they attributed divine significance to most numbers up to fifty.

One of the most absurd yet widely spread forms which numerlogy took was the so-called *Gematria*. Every letter in the Hebrew or Greek alphabet had the double meaning of a sound and of a number. The sum of the numbers represented by the letters of the word was the *number of the word*, and from the standpoint of Gematria two words were equivalent if they added up to the same number. Not only was Gematria used from the earliest days for the interpretation of Biblical passages, but there are indications that the writers of the Bible had practiced the art. Thus Abraham proceeding to the rescue of his brother Eliasar drives forth 318 slaves. Is it just a coincidence that the Hebrew word Eliasar adds up to 318?

Numerous examples of Gematria are found in Greek mythology. The names of the heroes Patroclus, Hector and Achilles add up to 87, 1225, and 1276 respectively. To this was attributed the superiority of Achilles. A poet desiring to confound his pet enemy, whose name was Thamagoras, proved that the word was equivalent to *loimos*, a sort of pestilence.

Christian theology made particular use of Gematria in interpreting the past as well as in forecasting the future. Of special significance was 666, the number of the Beast of Revelation. The Catholics' interpretation of the Beast was the Antichrist. One of their theologians, Peter Bungus, who lived in the days of Luther, wrote a book on numerology consisting of nearly 700 pages. A great part of this work was devoted to the mystical 666, which he had found equivalent to the name of Luther; this he took as conclusive evidence that Luther was the Antichrist.

In reply Luther interpreted 666 as the forecast of the duration of the Papal regime and rejoiced in the fact that it was so rapidly nearing its end.

Gematria is a part of the curriculum of the devout Hebrew scholar of today. How skilled these scholars are in this dual interpretation of Bibilical words is illustrated in this seemingly impossible feat. The Talmudist will offer to call out a series of numbers which follow no definite law of succession, some running as high as 500 and more. He will continue this perhaps for ten minutes, while his interlocutor is writing the numbers down. He will then offer to repeat the same numbers without an error and in the same succession. Has he memorized the series of numbers? No, he was simply translating some passage of the Hebrew scriptures into the language of Gematria.

But let us return to number worship. It found its supreme expression in the philosophy of the Pythagoreans. Even numbers they regarded as soluble, therefore ephemeral, feminine, pertaining to the earthly; odd numbers as indissoluble, masculine, partaking of celestial nature.

Each number was identified with some human attribute. *One* stood for reason, because it was unchangeable; *two* for opinion; *four* for justice, because it was the first perfect square, the product of equals; *five* for marriage, because it was the union of the first feminine and the first masculine number. (One was regarded not as an odd number, but rather as the *source* of all numbers.)

Strangely enough we find a striking correspondence in Chinese mythology. Here the odd numbers symbolized white, day, heat, sun, fire; the even numbers, on the other hand, black, night, cold, matter, water, earth. The numbers were arranged in a holy board, the Lo-Chou, which had magic properties when properly used.

"Bless us, divine number, thou who generatest gods and men! O holy, holy *tetraktys*, though that containest the root and the source of the eternally flowing creation! For the divine number begins with the profound, pure unity until it comes to the holy four; then it begets the mother of all, the all-compromising, the all-bounding, the first-born, the never-swerving, the never-tiring holy ten, the keyholder of all."

This is the prayer of the Pythagoreans addressed to the *tetraktys*, the holy fourfoldness, which was supposed to represent the four elements: fire, water, air and earth. The holy ten derives from the first four numbers by a union of 1, 2, 3, 4. There is the quaint story that Pythagoras commanded a new disciple to count to four:

"See what you thought to be four was really ten and a complete triangle and our password."

The reference to a complete triangle is important: it seems to indicate that in these early Greek days numbers were recorded by dots. In the accompanying figure the triangular numbers, 1, 3, 6, 10, 15, are shown as well as the square numbers; 1, 4, 9, 16, 25. As this was the actual beginning of number theory, this reliance on geometrical intuition is of great interest. The Pythagoreans knew that a square number of any rank is equal to the triangular number of the same rank increased by its predecessor. They proved it by segregating the dots and counting them, as shown in the figure. It is interesting to compare this method with one that a bright highschool boy would use today. The triangular number of rank n is obviously $1 + 2 + 3 + \ldots + n$, the sum of an arithmetic progression and equals $\frac{1}{2}n(n + 1)$.

TRIANGULAR AND SQUARE NUMBERS

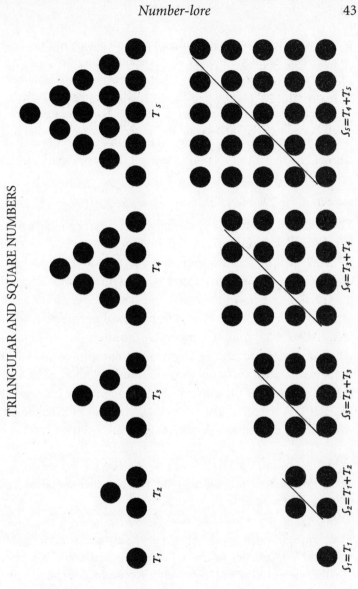

Its predecessor for the same reason is $\frac{1}{2}(n-1)n$. Simple algebra shows then that the two numbers added give n^2, the square number of rank n. (See figure, page 43.)

Today, nothing remains of this early geometric origin but the words *square* and *cube*. The triangular and more generally polygonal numbers are of little scientific interest. Yet even in the days of Nicomachus (100 A.D.) they were the principal objects of arithmetic investigation.

The source of this mystic philosophy of the Pythagoreans, which left such a deep impression on the speculations of all Greek thinkers including Plato and Aristotle, is still a controversial question. To the modern mind steeped in a rationalism the pompous number-worship may appear as *superstition erected into a system*. When we view it in historical perspective, we are inclined to take a more charitable attitude. Stripped of its religious mysticism, the Pythagorean philosophy contained the fundamental idea that only through number and form can man grasp the nature of the universe. Such thoughts are expressed by Philolaus, Pythagoras' ablest disciple, and also by Nicomachus, who may be classed as a neo-Pythagorean.

"All things which can be known have number; for it is not possible that without number anything can be either conceived or known." (*Philolaus.*)

"All things that have been arranged by nature according to a workmanlike plan appear, both individually and as a whole, as singled out and set in order by Foreknowledge and Reason, which created all according to Number, conceivable to mind only and therefore wholly immaterial; yet real; indeed, the really real, the eternal." (*Nicomachus.*)

Pythagoras, when asked what a friend was, replied: "One who is the other I, such are 220 and 284." Expressed in modern terminology this meant: the divisors of 284 are 1, 2, 4, 71, and 142, and these add up to 220; while the divisors of 220 are 1, 2, 4, 5, 10, 11, 20, 22, 44, 55, and 110, and these in turn add up to 284. Such numbers the Pythagoreans called *amicable* numbers.

The discovery of such couples was to the Greeks a problem of great interest and of considerable difficulty. The general question whether there exists an infinity of such couples has not been settled to this day, although almost a hundred are known.

The amicable numbers were known to the Hindus even before the days of Pythagoras. Also certain passages of the Bible seem to indicate that the Hebrews attached a good omen to such numbers.

There is an unauthenticated mediæval story of a prince whose name was from the standpoint of gematria equivalent to 284. He sought a bride whose name would represent 220, believing that this would be Heaven's guarantee of a happy marriage.

Then there were the *perfect* numbers. Consider first a number such as 14; add up its divisors which are 1, 2, and 7; we get 10. The number 14 therefore is greater than the sum of its own divisors, and is for this reason called *excessive*. On the other hand the sum of the divisors of 12 is 16—greater than 12, and for this reason 12 is said to be *defective*. But in a *perfect* number there is neither excess nor deficiency; the number equals the sum of its own divisors.

The smallest perfect numbers are 6 and 28, and were known to the Hindus as well as to the Hebrews. Some commentators of

the Bible regard 6 and 28 as the basic numbers of the Supreme
Architect. They point to the 6 days of creation and the 28 days of
the lunar cycle. Others go so far as to explain the imperfection of
the second creation by the fact that eight souls, not six, were res-
cued in Noah's ark.

Said St. Augustine:

> "Six is a number perfect in itself, and not because God created all
> things in six days; rather the converse is true; God created all
> things in six days because this number is perfect, and it would
> have been perfect even if the work of the six days did not exist."

The next two perfect numbers seem to have been the
discovery of Nicomachus. We quote from his *Arithmetica:*

> "But it happens that, just as the beautiful and the excellent are rare
> and easily counted but the ugly and the bad are prolific, so also
> excessive and defective numbers are found to be very many and in
> disorder, their discovery being unsystematic. But the perfect are
> both easily counted and drawn up in a fitting order: for only one is
> found in the units, 6; and only one in the tens, 28; and a third in the
> depth of the hundreds, 496; as a fourth the one, on the border of
> the thousands, that is short of the ten thousand, 8128. It is their
> uniform attribute to end in 6 or 8, and they are invariably even."

If Nicomachus meant to imply that there was a perfect number in every decimal class, he was wrong, for the fifth perfect
number is 33,550,336. But his guess was excellent in every other
respect. While the impossibility of an odd perfect was never
proved, no example of such a number is known. Furthermore it
is true that an even perfect must end in either 6 or 8.

How much importance the Greeks attached to the perfect
numbers is shown by the fact that Euclid devotes a chapter to
them in his *Elements*. He there proves that any number of the

form 2^{p-1} $(2^p - 1)$ is perfect, provided the odd factor, $2^p - 1$, is prime. Until recently only twelve numbers were definitely known to satisfy these conditions. The values of the exponent p for these perfect numbers are

$$p = 2, 3, 5, 7, 13, 17, 19, 61, 107, 127, 257.$$

With the advent of the new high-speed calculating machines, five more have been added to this list.

Prime numbers were a subject of great interest from the earliest days. Of the various methods used the most interesting was one known as the *Sieve,* which is attributed to Eratosthenes, a contemporary of Archimedes. The sieve of Eratosthenes for the first 100 numbers is shown in the accompanying figure. The scheme consists in writing down all the integers in their natural succession and then striking out first all the multiples of 2, then the remaining multiples of 3, then those of 5, etc. If we want to determine all the prime numbers less than a thousand, for instance, it is not necessary to go beyond the multiples of 31, because $31^2 = 961$ is the largest square of a prime number less than 1,000. In a somewhat modified form, this scheme is used today for the construction of tables of prime numbers. Modern tables extend as far as 10,000,000.

Ingenious as this method of elimination is, it is purely inductive and therefore incapable of proving general properties of prime numbers. For instance, the first question which naturally arises is whether the primes form a finite or an infinite collection. In other words, is the number of primes unlimited, or is there a greatest prime? Of this fundamental problem Euclid gave a solution which is inscribed in the annals of mathematics as a model of perfection.

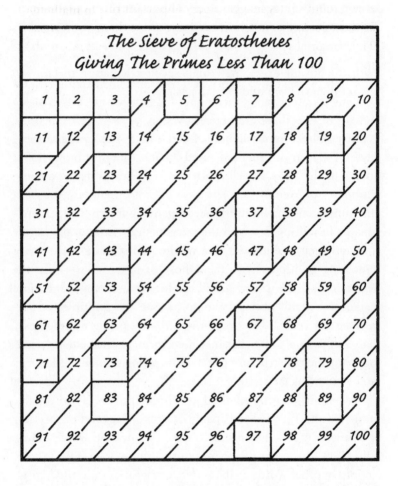

The Sieve of Eratosthenes
Giving The Primes Less Than 100

In this proof Euclid introduces for the first time in history what we call today *factorial numbers*. These products of the first *n* consecutive integers play a very important rôle in mathematical questions. The notation used to express these numbers is *n!*. Thus factorial seven is $7! = 1 \cdot 2 \cdot 3 \cdot 4 \cdot 5 \cdot 6 \cdot 7$. A table of factorials up to 11! is given on page 53.

To prove that there is no greatest prime, Euclid shows that if *n* is any prime number, then either the number $(n! + 1)$ is also a prime, or else there are between *n* and $(n! + 1)$ other primes. Both cases are possible: thus when *n* is 3 the corresponding Euclid number is 7, a prime number; but when *n* is 7, $n! = 5040$, and the corresponding Euclid number is 5041, a composite number, in fact the perfect square, 71×71. Between 7 and $(7! + 1)$ there is, therefore, the prime number 71.

To prove this in the general case, Euclid proceeds essentially as follows: Two consecutive numbers can have no divisors in common; this is particularly true of *n!* and $(n! + 1)$; if then the latter possesses any prime divisors at all, these must be distinct from *n* or any number preceding *n*. Either then the Euclid number $(n! + 1)$ contains a prime divisor greater than *n*, or else the Euclid number itself is prime: in either case there are primes greater than *n*.

He concludes that there cannot be a greatest prime, which is another way of saying that the number of primes is infinite.

The next question concerns the distribution of primes. We may speak, for instance, of the *density* of primes, i.e., the number of primes in, say, any thousand numbers. This is, of course, the same as enumerating the number of primes which are less than a given number. The greatest ingenuity has been displayed by many modern mathematicians in attacking this problem, but a wholly satisfactory solution has not yet been attained. However, we know enough to conclude that the primes do not become substantially rarer as we go on.

In 1845, the French mathematician Bertrand asserted that between any number and its double there exists at least one prime. He based this assertion on an empirical study of a table of primes. For over fifty years this proposition was known as the *postulate of Bertrand*. It was finally proved by the great Russian mathematician Tchebyshev who has also shown that there are primes even between much narrower limits. Finally in 1911 the Italian mathematician Bonolis advanced this problem considerably by giving an approximate formula for the number of primes between x and $\frac{3}{2}x$. According to this formula there are not fewer than a million primes between 100,000,000 and 150,000,000.

On the other hand it has been shown that the so-called twin-primes, such as (3,5), (5,7), (11,13), (17,19), (29,31), (41,43), etc., become rarer and rarer as the numbers increase. This remarkable theorem was proved by the Dutch mathematician Bruns in 1919.

How do we recognize whether a given number is prime or composite? We know that the number is composite if it ends in a 5, or in a 0, or in an *even* digit. But suppose it ends in a 3, 7, or 9. Then we have a relatively simple test for the divisibility by 3 or 9: if the sum of all digits is a multiple of 3 or 9 then the number itself is also a multiple of 3 or 9. This so-called *rule of nine is* very old.

For other divisors the conditions are much more involved. It is true that Pascal in 1654 and Lagrange a hundred years later established very general theorems, but these were distinguished rather by mathematical elegance than practical value. Professor

Dickson in his *History of the Theory of Numbers* makes the following characteristic remark:

> "To tell whether a given number of 15 or 20 digits is prime or not, all time would not suffice for the test, whatever use is made of what is already known."

It is not surprising then that for centuries all sorts of attempts were made to find a general mathematical formula which would fit all primes, or, failing in this, at least some particular scheme to generate primes. In 1640 the great French mathematician Fermat announced that he had found a form which represented primes only. The numbers so generated are today called Fermat numbers.

Here are the first four Fermat numbers:

$$2^2 + 1 = 5; \; 2^{2^2} + 1 = 17; \; 2^{2^3} + 1 = 257; \; 2^{2^4} + 1 = 65537.$$

Fermat checked the primality of these first numbers and was for a while convinced of the generality of his theorem. Later, however, he began to doubt. As a matter of fact about a hundred years later, Euler showed that the fifth Fermat number was already composite, one of its factors being 641. Since then the same fact has been established for the sixth, seventh, and a dozen higher Fermat numbers.

This illustrates the danger of *incomplete* induction. Still more striking examples are furnished by some quadratic expressions, such as

$$f(n) = n^2 - n + 41$$

We find upon substitution: $f(1) = 41, f(2) = 43, f(3) = 47, f(4) = 53$ ... which are all prime, and this fact may be verified up to $n = 40$. But for $n = 41$, we obviously get a composite number: $f(41) = 41 \times 41$.

The failure to obtain a general form for generating prime numbers led to indirect criteria for testing primality. Fermat thought that he had found such a criterion in the theorem: *n being any integer whatsoever, the binomial $n^p - n$ is a multiple of p, if p is prime.* As an illustration let us consider the case $p = 5$. We have

$$n^5 - n = n(n^4 - 1) = n(n^2 + 1)(n^2 - 1)$$

and it is readily seen that one of three factors must be a multiple of five, no matter what n may be.

The truth of the Fermat theorem was established by Leibnitz, Euler and others. One trouble with it is that while it is true, it is not a *criterion*, i.e., *the condition is necessary but not sufficient.* For instance 341 is not a prime, yet $2^{341} - 2$ contains 341 as a factor.

A criterion, that is, a condition which is both necessary and sufficient, is furnished by the so called *Wilson Theorem*. It would be more just to name it after Leibnitz, who was the first to prove that the condition is necessary. A hundred years later Lagrange showed that the condition is also sufficient. Consider the accompanying table of factorials, and their successors, the Euclid numbers, $n! + 1$. Observe that when $n + 1$ equals 2, 3, 5, 7, 11, which are all primes, $(n + 1)$ is a divisor of $(n! + 1)$; while for the composite values of $(n + 1)$, such as 4, 6, 8, 9, 10, the division of $n! + 1$ by $n + 1$ will leave a remainder. This is a perfectly general property: *The condition necessary and sufficient that p be a prime is that the successor of $(p - 1)!$ contain p as a factor.*

This remarkable proposition is of great theoretical interest. However, a direct verification as to whether $(p - 1)! + 1$ has p for

The Wilson Criterion

The Wilson index is the remainder of divison of $(n-1)! + 1$ by n. If the Wilson index is 0, the number is prime.

n	Character	Factorials $(n-1)!$		$(n-1)! + 1$ The Euclid Numbers	The Wilson Index
2	Prime	$1! =$	1	2	0
3	Prime	$2! =$	2	3	0
4	Composite	$3! =$	6	7	3
5	Prime	$4! =$	24	25	0
6	Composite	$5! =$	120	121	1
7	Prime	$6! =$	720	721	0
8	Composite	$7! =$	5,040	5,041	1
9	Composite	$8! =$	40,320	40,321	1
10	Composite	$9! =$	362,880	362,881	1
11	Prime	$10! =$	3,628,800	3,628,801	0
12	Composite	$11! =$	39,916,800	39,916,801	1

a divisor is as difficult as testing directly whether p is prime, when p is a large number.

Many indirect propositions have been established since. One of the most interesting is the *postulate of Goldbach*, a contemporary of Euler. This postulate alleges that *every even number is the sum of two primes*. This postulate has been verified for all numbers up to 10,000, and some beyond. But the proof of this important allegation is still challenging the ingenuity of mathematicians.

"It is impossible to partition a cube into two cubes, or a biquadrate into two biquadrates, or generally any power higher than a square into two powers of like degree. I have discovered a truly wonderful proof of this, which, however, this margin is too narrow to hold."

This famous marginal note will soon be three hundred years old, and many a mathematician has since wished that Fermat had had a wider margin at his disposal when he wrote it.

The history of this problem goes back to Egyptians, who knew of the existence of a right triangle, which has its three sides in the ratio 3:4:5. In fact they used this triangle as a sort of carpenter's square. I understand that the Chinese use some such scheme even today.

| The Pythagorean Numbers | | | | | $x = u + \sqrt{2uv}$ | | |
| For any value of u and v such that 2uv is a perfect square we obtain a Pythagorean triangle. | | | | | $y = v + \sqrt{2uv}$ $z = u + v + \sqrt{2uv}$ | | |
2uv	uv	$\sqrt{2uv}$	u	v	x	y	z
4	2	2	1	2	3	4	5
16	8	4	1	8	5	12	13
16	8	4	2	4	6	8	10
36	18	6	1	18	7	24	25
36	18	6	2	9	8	15	17
36	18	6	3	6	9	12	15
64	32	8	1	32	9	40	41
64	32	8	2	16	10	24	26

Is this the only right triangle the sides of which can be expressed as integers? No, there is an infinite number of other such triplets, quite a few of which were known to the Pythagoreans. Diophantus of Alexandria, who lived in the third century of our era, gave in his *Arithmetica* a rule for determining such numbers. In modern notation this problem is equivalent to the solution of the equation

$$x^2 + y^2 = z^2$$

in whole numbers. A few of these Pythagorean numbers are given in the accompanying table, and the formula heading the table enables us to derive "all" Pythagorean numbers. It is obvious from this formula that not only does the equation $x^2 + y^2 = z^2$ admit integers as solutions, but that there exists an infinity of such solutions.

It was natural to inquire whether the same is true of similar equations of higher degree.

A new edition of Diophantus's *Arithmetica* appeared in France in about 1621, and a copy came into the possession of Fermat. On one of the pages of this book Fermat wrote the marginal note which has puzzled the mathematical world ever since. In modern terminology Fermat's statement can be formulated in the following way: *show that the equation*

$$x^n + y^n = z^n,$$

where x, y and z are to be integers, is impossible when n is an integer greater than two.

What is the present status of this problem? Euler showed the impossibility when *n* is 3 or 4; Dirichlet proved it for *n* = 5. It has been demonstrated that if the proposition were true for prime values of the exponent *n*, it would be true for composite exponents. It has been established that the statement is true for certain very numerous forms of the exponent, and that the Fermat equation has no solution if *n* is less than 269. Yet, the general proposition has not been proved, and it may be seriously doubted whether Fermat had a general proof of his theorem.

The Fermat problem came in for a good deal of publicity because of the sensational announcement that a prize of 100,000 marks would be awarded for its complete solution. This fund was bequeathed in 1908 by a Dr. Wolfskoel who had himself devoted considerable time to the problem, without advancing it, however. Since then many amateurs who had hitherto directed their energy to such problems as the *squaring of the circle, the trisection of an angle,* or the invention of *perpetual-motion machines,* have begun to concentrate on the Fermat theorem.

It is estimated that over a thousand such "complete" solutions reached the committee on award between 1908 and 1911. Luckily the announcement stipulated that contributions must be printed, and this may have dampened the ardor of many. We note with interest that most of the "solutions" submitted were published by the authors themselves. It is characteristic of all such efforts that their authors completely ignore the tremendous amount of work already accomplished; nor are they interested in learning wherein the difficulty lies.

The problem attracted the attention of the greatest mathematicians of the last three centuries: Euler and Lagrange, Kummer and Riemann, all tried in vain to prove or disprove it. If all the articles published on this and related subjects were gathered together they would fill a small library.

Out of the attempts to solve the Fermat problem grew up a science which far transcends in importance the original problem. So important and far-reaching are some of these results that we may consider it fortunate that the original problem has not been solved. Thus, while trying to prove Fermat's theorem, Eduard Kummer created his famous theory of *ideal numbers,* one of the most fundamental and most fertile achievements of the nineteenth century. However, the scope of this book does not permit me to give even a summary description of this far-reaching conception.

Born in religious mysticism, the theory of integers passed through a period of erratic puzzle-solving before it acquired the status of a science.

Paradoxical though it may seem to those who identify the mystic with the abstract, the basis of this number mysticism was concrete enough. It revolved about two ideas. The figurative numbers of the Pythagoreans, of very ancient origin, show the close connection between *form* and *number*. Numbers which represented simple and regular figures, like triangles, squares, pyramids, and cubes, were easier to conceive and were therefore singled out as of special importance. On the other hand, we have the perfects, the amicables, and the primes, which have special properties with respect to divisibility. These can be traced to the importance the ancients attached to problems of *distribution*, as the Sumerian clay tablets and the earliest Egyptian papyri clearly show.

This concreteness accounts for the experimental character of the early period, a character which the theory has to a degree preserved up to now. I leave the word to one of the most eminent number theorists of our day, the late G.H. Hardy of England.

"The theory of numbers, more than any other branch of mathematics, began by being an experimental science. Its most famous theorems have all been conjectured, sometimes a hundred years or more before they were proved; and they have been suggested by the evidence of a mass of computations."

The concrete has ever preceded the abstract. That is why the theory of numbers preceded arithmetic. And the concrete has ever been the greatest stumbling block to the development of a science. The peculiar fascination which *numbers as individuals* have exerted on the mind of man from time immemorial was the main obstacle in the way of developing a *collective* theory of numbers, i.e., an arithmetic; just as the concrete interest in individual stars long delayed the creating of a scientific astronomy.

CHAPTER 4

The Last Number

*"But what has been said once, can always
be repeated."*
—Zeno of Elea, as quoted by Simplicius

What is there in mathematics that makes it the acknowledged model of the sciences called exact, and the ideal of the newer sciences which have not yet achieved this distinction? It is, indeed, the avowed ambition of the younger investigators at least, in such fields as biology or the social sciences, to develop standards and methods which will permit these to join the ever-growing ranks of sciences which have already accepted the domination of mathematics.

Mathematics is not only the model along the lines of which the exact sciences are striving to design their structure; mathematics is the cement which holds this structure together. A problem, in fact, is not considered solved until the studied phenomenon has been formulated as a mathematical law. Why is it believed that only mathematical processes can lend to observation, experiment, and speculation that precision, that conciseness, that solid certainty which the exact sciences demand?

When we analyze these mathematical processes we find that they rest on the two concepts: Number and Function; that Function itself can in the ultimate be reduced to Number; that the general concept of Number rests in turn on the properties we ascribe to the natural sequence: one, two, three ....

It is then in the properties of the whole numbers that we may hope to find the clue to this implicit faith in the infallibility of mathematical reasoning!

The first practical application of these properties take the form of the elementary operations of arithmetic; *addition, subtraction, multiplication,* and *division* of whole numbers. We learn these operations very early in life and it is not surprising that most of us have completely forgotten the circumstances under which we acquired them. Let us refresh our memory.

We began by memorizing the table $1 + 1 = 2, 1 + 2 = 3, \ldots$ We were drilled and drilled until we were able to add up without hesitancy any two numbers up to ten. In the course of this first phase of our instruction, we were taught to observe that $5 + 3 = 3 + 5$ and that this was not an accident, but a general rule. Later we learned to express this property of addition in words: the *sum does not depend on the order of its terms.* The mathematician says no more when he states: *addition is a commutative operation,* and writes in symbols:

$$a + b = b + a.$$

We were next shown that $(2 + 3) + 4 = 2 + (3 + 4)$; by this was meant that whereas $(2 + 3) + 4$ meant that we add 3 to 2 and 4 to the sum, it was really immaterial in what order we added, for the same result would be obtained if to 2 were added the sum of $(3 + 4)$. The mathematician says no more when he states that addition is an *associative* operation, and writes

$$(a + b) + c = a + (b + c)$$

We never attached much importance to these statements. Yet they are fundamental. On them is based the rule for adding larger numbers. The scheme

$$
\begin{array}{r}
25 \\
34 \\
\underline{56} \\
115
\end{array}
$$

is but a compact paraphrase of:

$$25 + 34 + 56 = (20 + 5) + (30 + 4) + (50 + 6) =$$
$$(20 + 30 + 50) + (5 + 4 + 6) = 100 + 15 = 115$$

in which the commutativity and associativity of addition play a fundamental rôle.

We then proceeded to *multiplication*. Again we memorized a long table until we could tell mechanically the product of any two numbers up to ten. We observed that like addition, *multiplication was both associative and commutative*. Not that we used these words, but we implied as much.

There was yet another property which concerned multiplication and addition jointly. The product $7 \times (2 + 3)$ means that seven is to be multiplied by the sum $(2 + 3)$, that is, by 5; but the same result could be obtained by adding the two partial products (7×2) and (7×3). The mathematician expresses this in the general statement: multiplication is *distributive* with respect to addition, and writes

$$a(b + c) = ab + ac.$$

It is this distributivity which is at the bottom of the scheme which we use in multiplying numbers greater than ten. Indeed, when we analyze the operation

$$
\begin{array}{r}
25 \\
\underline{43} \\
75 \\
\underline{100} \\
1075
\end{array}
$$

we find it but a compact paraphrase of the involved chain of operations in which this distributive property is freely used.

Thus

$$25 \times 43 = (20 + 5) \times (40 + 3) = [(20 + 5) \times 3] + [(20 + 5) \times 40] = (20 \times 3) + (5 \times 3) + (20 \times 40) + (5 \times 40) = 75 + 1000 = 1075$$

Such are the facts which form the basis of the mathematical education of all thinking men, nay of all people who have had any schooling at all. On these facts is built *arithmetic*, the foundation of mathematics, which in turn supports all science pure and applied which in turn is the fertile source of all technical progress.

Later new facts, new ideas, new concepts were added to our mental equipment, but none of these had to our mind the same security, the same solid foundation, as these properties of whole numbers, which we acquired at the tender age of six. This is expressed in the popular saying: It is as obvious as that two and two make four.

We learned these at an age when we were interested in the "how" of things. By the time we were old enough to ask "why," these rules, through constant use, had become such an intimate part of our mental equipment that they were taken for granted.

The individual is supposed to have retraced in his development the evolution of the species to which he belongs. Some such principle governs the growth of the human intellect as well. In the history of mathematics, the "how" always preceded the "why," the technique of the subject preceded its philosophy.

This is particularly true of arithmetic. The counting technique and the rules of reckoning were established facts at the end of the Renaissance period. But the philosophy of number did not come into its own until the last quarter of the nineteenth century.

As we grow older, we find ample opportunity to apply these rules in our daily tasks, and we grow more and more confident of their generality. The strength of arithmetic lies in its *absolute generality*. Its rules admit of no exceptions: they apply to *all numbers*.

All numbers! Everything hangs on this short but so tremendously important word *all*.

There is no mystery about this word, when it is applied to any *finite* class of things or circumstances. When, for instance, we say "all living men," we attach a very definite meaning to it. We can imagine all mankind arranged in an array of some sort: in this array there will be a *first* man, and there will be a *last* man. To be sure, to prove in all rigor a property true of all living men we should prove it for each individual. While we realize that the actual task would involve insurmountable difficulties, these difficulties, we feel, are of a purely *technical* and not of a *conceptual* character. And this is true of any *finite* collection, i.e., of any collection which has a *last* as well as a *first* member, for a*ny such collection can be exhausted by counting.*

Can we mean the same thing when we say *all numbers?* Here too, the collection can be conceived as an array, and this array will have a first member, the number *one*. But how about the last?

The answer is ready: *There is no last number!* The process of counting cannot conceivably be terminated. *Every number has a successor.* There is an *infinity* of numbers.

But if there be no last number, what do we mean by all numbers, and particularly, what do we mean by *the property of all numbers?* How can we prove such a property: certainly not by testing every individual case, since we know beforehand that we cannot possibly exhaust all cases.

At the very threshold of mathematics we find this *dilemma of infinity,* like the legendary dragon guarding the entrance to the enchanted garden.

What is the source of this concept of infinity, this faith in the inexhaustibility of the counting process? Is it experience? Certainly not! Experience teaches us the finitude of all things, of all human processes. We know that any attempt on our part to exhaust number by counting would only end in our own exhaustion.

Nor can the existence of the infinite be established mathematically, because infinity, the inexhaustibility of the counting process, is a mathematical assumption, *the basic assumption of arithmetic,* on which all mathematics rests. Is it then a supernatural truth, one of those few gifts which the Creator bestowed upon man when he cast him into the universe, naked and ignorant, but free to shift for himself? Or has the concept of infinity grown upon man, grown out, indeed, out of his futile attempts to reach the last number? Is it but a confession of man's impotence to exhaust the universe by number?

"There is a last number, but it is not in the province of man to reach it, for it belongs to the gods." Such is the keynote of most ancient religions. The stars in the heavens, the grains of sand, the drops of the ocean exemplify this *ultra-ultimate* which is beyond the mind of man to reach. "He counted the stars and named them all," says the psalmist of Jehovah. And Moses in invoking the promise of God to his chosen people says: "He who can count the dust of the earth will also count your seed."

"There are some, King Gelon, who think that the number of the sands is infinite in multitude; and I mean by sand not only that which exists about Syracuse and the rest of Sicily but also that which is found in every region whether inhabited or uninhabited. Again there are some who, without regarding it as infinite, yet think that no number has been named which is great enough to exceed its multitude. And it is clear that they who hold this view, if they imagined a mass made up of sand in other respects as large as the mass of the earth, including in it all the seas and the hollows of the earth filled up to the height equal to that of the highest mountains, would be many times further still from recognizing that any number could be expressed which exceeded the multitude of the sand so taken. But I will try to show you, by means of geometrical proofs which you will be able to follow, that, of the numbers named by me and given in the work which I sent to Zeuxippus, some exceed not only the number of the mass of sand equal in size to the earth filled up in the way described, but also that of a mass equal in size to the universe."

(Archimedes: The Sand Reckoner)

Now this universe of Archimedes was a sphere limited by the fixed stars. This sphere he estimated to be of a diameter equal to 10,000 earth-diameters. Assuming the number of grains of sand which would fill a poppy seed as 10,000, and the diameter of the earth not greater than 10,000 miles (300,000 stadia), he found for the grains of sand that would fill the universe a fabulous number which in our numeration would be expressed in 52 digits. To express this number Archimedes invented a new unit, *the octade*, which corresponded to our 100,000.

The history of the attempts to square the circle will furnish another example. The problem in its original form was to construct by ruler and compass a square of an area equal to that of a given circle. Now, it is possible to construct a square equivalent to an inscribed regular polygon of say 8 sides. On the other hand it is observed that if we increase the number of sides to 16, 32, 64, etc. we shall approximate the area of the circle more and more closely. Now there is no doubt that some of the Greek

geometers regarded this doubling process not as an approximation but as a means of attaining the circle, i.e., they thought if
they could continue this process long enough they would eventually reach the ultimate polygon which would coincide with the
circle at all points.

It is a plausible hypothesis that the early conception of
infinity was not the uncountable, but the yet-uncounted. The
last number meant *patience* and *perseverance*, and man seemed
to be lacking in these qualities. It was of the same order of things
as reaching heaven in the story of the Tower of Babel. The last
number, like the heavens, belonged to God. In His jealous wrath
He would confound the tongues of the ambitious builders.

This confusion of tongues persists to this day. Around infinity
have grown up all the paradoxes of mathematics: from the arguments of Zeno to the antimonies of Kant and Cantor. This story
we shall tell in another chapter. What concerns us here is that
these paradoxes were instrumental in creating a more critical
attitude towards the foundations of arithmetic. For, since the
properties of whole numbers form the basis of mathematics, if
these properties can be proved by the rules of *formal* logic, then
all of mathematics is a logical discipline. If, however, logic is
insufficient to establish these properties, then mathematics is
founded on something more than mere logic: its creative power
relies on that elusive, intangible thing which is called human
intuition.

Let there be no misunderstanding! It is not the validity of
these properties of number which is at stake; the issue is the
validity of the arguments which purport to prove the validity of
these properties. The questions that have been at issue ever since

the foundations of mathematics were submitted to this searching analysis, the questions which have split the leading mathematical thinkers into two contending camps, *intuitionists* vs. *formalists,* are these: What constitutes a mathematical proof? What is the nature of reasoning generally and mathematical reasoning in particular? What is meant by *mathematical existence?*

Now, the laws of sound reasoning are as old as the hills. They were formulated in a systematic manner by Aristotle, but were known long before him. Why, they are the very skeleton of the human intellect: every intelligent man has occasion to apply these laws in his daily pursuits. He knows, that in order to reason soundly, he must first define his premises without ambiguity, then through a step-by-step application of the canons of logic he will eventually arrive at a conclusion which is the *unique* consequence of the logical process he used in reaching it.

If this conclusion does not tally with the facts as we observe them, then the first step is to find out whether we applied these canons correctly. This is not the place to analyze the validity of these canons. Not that they have been spared the scorching fire of this critical age! Quite the contrary: one of them is, indeed, the center of a controversy which has been raging for a quarter of a century and which shows no sign of abating. However, this is a story by itself and it will be told in its proper place.

If it is found that the canons of logic were applied correctly, then the discrepancy, if there be a discrepancy, may mean that there is something wrong with our premises. There may be an inconsistency lurking somewhere in our assumptions, or one of our premises may contradict another.

Now, to establish a set of assumptions for any particular body of knowledge is not an easy task. It requires not only acute analytical judgment, but great skill as well. For, in addition to this freedom from contradiction, it is desired that each assumption should be independent of all the others, and that the whole system be exhaustive, i.e., completely cover the question under investigation. The branch of mathematics which deals with such problems is called *axiomatics* and has been cultivated by such men as Peano, Russell and Hilbert. In this manner logic, formerly a branch of philosophy, is being gradually absorbed into the body of mathematics.

Returning to our problem, suppose that we have examined our premises and have found them free from contradictions. Then we say that our conclusion is logically flawless. If, however, this conclusion does not agree with the observed facts, we know that the assumptions we have made do not fit the concrete problem to which they were applied. There is nothing wrong with the tailoring of the suit. If it bulges in some spots and cracks in others, it is the fault of the *fitter*.

The process of reasoning just described is called *deductive*. It consists in starting from very general properties, which take the form of *definitions, postulates* or *axioms,* and in deriving from these, by means of the canons of logic, statements concerning things or circumstances which would occur in particular instances.

The process of deduction is characteristic of mathematical reasoning. It has found a nearly complete realization in geometry, and for this reason the logical structure of geometry has been the model for all exact sciences.

Quite different in its nature is the other method used in scientific investigation: *induction.* It is generally described as proceeding from the particular to the general. It is the result of observation and experience. To discover a property of a certain class of objects we repeat the observation or tests as many times as feasible, and under circumstances as nearly similar as possible. Then it may happen that a certain definite tendency will manifest itself throughout our observation or experimentation. This tendency is then accepted as the property of the class. For example, if we subject a sufficiently large number of samples of lead to the action of heat, and we find that in every case melting began when the thermometer reached 328°, we conclude that the point of fusion of lead is 328°. Back of this is the conviction that no matter how many more samples we might test, the circumstances not having changed, the results would also be the same.

This process of induction, which is basic in all experimental sciences, is *for ever banned* from rigorous mathematics. Not only would such a proof of a mathematical proposition be considered ridiculous, but even as a verification of an established truth it would be inacceptable. *For, in order to prove a mathematical proposition, the evidence of any number of cases would be insufficient, whereas to disprove a statement one example will suffice.* A mathematical proposition is true, if it leads to no logical contradiction, false otherwise. *The method of deduction is based on the principle of contradiction and on nothing else.*

Induction is barred from mathematics and for a good reason. Consider the quadratic expression $(n^2 - n + 41)$ which I mentioned in the preceding chapter. We set in this expression $n = 1, 2, 3 \ldots\ldots$ up to $n = 40$: in each of these cases we get a prime number as the result. Shall we conclude that this expression represents prime numbers for all values of n? Even the least mathematically trained reader will recognize the fallacy of such a conclusion: yet many a physical law has been held valid on less evidence.

Mathematics is a deductive science, arithmetic is a branch of mathematics. Induction is inadmissible. The propositions of arithmetic, the associative, commutative and distributive properties of the operations, for instance, which play such a fundamental rôle even in the most simple calculations, must be demonstrated by deductive methods. What is the principle involved?

Well, this principle has been variously called *mathematical induction,* and *complete induction,* and that of *reasoning by recurrence.* The latter is the only acceptable name, the others being misnomers. The term induction conveys an entirely erroneous idea of the method, for it does not imply systematic trials.

To give an illustration from a familiar field, let us imagine a line of soldiers. Each one is instructed to convey any information that he may have obtained to his neighbor on the right. The commanding officer who has just entered the field wants to ascertain whether *all* the soldiers know of a certain event that has happened. Must he inquire of every soldier? Not if he is sure that whatever any soldier may know his neighbor to the right is also bound to know, for then if he has ascertained that the *first* soldier to the left knows of the event he can conclude that *all* the soldiers know of it.

The argument used here is an example of reasoning by recurrence. It involves two stages. It is first shown that the proposition we wish to demonstrate is of the type which Bertrand Russell calls *hereditary:* i.e., if the proposition were true for any member of a sequence, its truth for the *successor* of the member would follow as a logical necessity. In the second place, it is shown that the proposition is true for the first term of the sequence. This latter is the so-called *induction* step. Now in view of its hereditary nature, the proposition, being true of the first term, must be true of the second, and being true of the second it must be true of the third, etc., etc. We continue in this way till we have exhausted the whole sequence, i.e., reached its *last* member.

Both steps in the proof, the induction and the hereditary feature, are necessary; neither is sufficient alone. The history of the two theorems of Fermat may serve as illustration. The first theorem concerns the statement that $2^{2^n} + 1$ is a prime for all values of n. Fermat showed by actual trial that such is the case for $n = 0, 1, 2, 3$ or 4. But he could not prove the hereditary property; and as a matter of fact, we saw that Euler disproved the proposition by showing that it fails for $n = 5$. The second theorem alleges that the equation $x^n + y^n = z^n$ cannot be solved in integers when n is greater than 2. Here the induction step would consist in showing that the proposition holds for $n = 3$, i.e., that the equation $x^3 + y^3 = z^3$ cannot be solved in whole numbers. It is possible that Fermat had a proof of this, and if so, here would be one interpretation of the famous marginal note. At any rate, this first step, we saw, was achieved by Euler. It remains to show that the property is hereditary, i.e., assuming it true for some value of n, say p, it should follow as a logical necessity that the equation $x^{p+1} + y^{p+1} = z^{p+1}$ cannot be solved in integers.

It is significant that we owe the first explicit formulation of the *principle of recurrence* to the genius of Blaise Pascal, a contemporary and friend of Fermat. Pascal stated the principle in a tract called *The Arithmetic Triangle* which appeared in 1654. Yet it was later discovered that the gist of this tract was contained in the correspondence between Pascal and Fermat regarding a problem in gambling, the same correspondence which is now regarded as the nucleus from which developed the theory of probabilities.

It surely is a fitting subject for mystic contemplation, that the principle of reasoning by recurrence, which is so basic in pure mathematics, and the theory of probabilities, which is the basis of all inductive sciences, were both conceived while devising a scheme for the division of the stakes in an unfinished match of two gamblers.

How the principle of mathematical induction applies to Arithmetic can be best illustrated in the proof that addition of whole numbers is an *associative* operation. In symbols this means:

(1) $$a + (b + c) = (a + b) + c$$

Let us analyze the operation $a + b$: it means that to the number a was added 1, to the result was added 1 again, and that this process was performed b times. Similarly $a + (b + 1)$ means $b + 1$ successive additions of 1 to a. It follows therefore that:

(2) $$a + (b + 1) = (a + b) + 1$$

and this is proposition (1) for the case when $c = 1$. What we have done, so far, constitutes the *induction* step of our proof.

Now for the hereditary feature. Let us assume that the proposition is true for some value of c, say n, i.e.

(3) $$a + (b + n) = (a + b) + n$$

Adding 1 to both sides:

(4) $$[a + (b + n)] + 1 = [(a + b) + n] + 1$$

which because of (2) can be written as

(5) $$(a + b) + (n + 1) = a + [(b + n) + 1]$$

And for the same reason is equivalent to

(6) $$(a + b) + (n + 1) = a + [b + (n + 1)]$$

but this is proposition (1) for the case $c = n + 1$.

Thus the fact that the proposition is true for some number *n* carries with it *as a logical necessity* that it must be true for the successor of that number, *n* + 1. Being true for 1, it is therefore true for 2; being true for 2, it is true for 3; and so on *indefinitely*.

The principle of mathematical induction in the more general form in which it is here applied can be formulated as follows: Knowing that a proposition involving a sequence is true for the first number of the sequence, and that the assumption of its truth for some particular member of the sequence involves as a logical consequence the truth of the proposition for the successor of the number, we conclude that it is true for all the numbers of the sequence. The difference between the *restricted* principle as it was used in the case of the soldiers, and the *general* principle as it is used in arithmetic, is merely in the interpretation of the word *all*.

Let me repeat: it is not by means of the restricted, but of the general principle of mathematical induction that the validity of the operations of arithmetic which we took on faith when we were first initiated into the mysteries of number has been established.

The excerpts in the following section are taken from an article by Henri Poincaré entitled *The Nature of Mathematical Reasoning*. This epoch-making essay appeared in 1894 as the first of a series of investigations into the foundations of the exact sciences. It was a signal for a throng of other mathematicians to inaugurate a movement for the revision of the classical concepts, a movement which culminated in the nearly complete absorption of logic into the body of mathematics.

The great authority of Poincaré, the beauty of his style, and the daring iconoclasm of his ideas carried his work far beyond the limited public of mathematicians. Some of his biographers estimated that his writing reached half a million people, an

audience which no mathematician before him had ever commanded.

Himself a creator in practically every branch of mathematics, physics, and celestial mechanics, he was endowed with a
tremendous power of introspection which enabled him to analyze the sources of his own achievements. His penetrating mind
was particularly interested in the most elementary concepts,
concepts which the thick crust of human habit has made almost
impenetrable: to these concepts belong *number, space,* and *time.*

"The very possibility of a science of mathematics seems an insoluble contradiction. If this science is deductive only in appearance, whence does it derive that perfect rigor which no one dares
to doubt? If, on the contrary, all the propositions it enunciates
can be deduced one from the other by the rules of formal logic,
why is not mathematics reduced to an immense tautology? The
syllogism can teach us nothing that is essentially new, and, if
everything is to spring from the principle of identity, everything
should be capable of being reduced to it. Shall we then admit that
the theorems which fill so many volumes are nothing but devi-
ous ways of saying that A is A?

"We can, no doubt, fall back on the axioms, which are the
source of all these reasonings. If we decide that these cannot be
reduced to the principle of contradiction, if still less we see in
them experimental facts, ... we have yet the resource of regarding them as *a priori* judgments. This will not solve the difficulty
but only christen it ....

"The rule of reasoning by recurrence is not reducible to the
principle of contradiction. ... Nor can this rule come to us from
experience. Experience could teach us that the rule is true for the
first ten or hundred numbers; it cannot attain the indefinite
series of numbers, but only a portion of this series, more or less
long, but always limited.

"Now, if it were only a question of a portion, the principle of contradiction would suffice; it would always allow of our developing as many syllogisms as we wished. It is only when it is a question of including an infinity of them in a single formula, it is only before the infinite, that this principle of logic fails, and here is where experience too becomes powerless ....

"Why then does this judgment force itself upon us with such an irresistible force? It is because it is only the affirmation of the power of the mind which knows itself capable of conceiving the indefinite repetition of the same act when this act is possible at all ....

"There is, we must admit, a striking analogy between this and the usual procedure of induction. But there is an essential difference. Induction, as applied in the physical sciences, is always uncertain, because it rests on the belief in a general order in the universe, an order outside of us. On the contrary, mathematical induction, i.e., demonstration by recurrence, imposes itself as a necessity, because it is only a property of the mind itself ....

"We can ascend only by mathematical induction, which alone can teach us something new. Without the aid of this induction, different from physical induction but just as fertile, deduction would be powerless to create a science.

"Observe, finally, that this induction is possible only if the same operation can be repeated indefinitely. That is why the theory of chess can never become a science: the different moves of the game do not resemble one another."

The last word should go to the master and so I should have liked to conclude this chapter. But history is no respecter of persons: the ideas of Poincaré raised a controversy which rages to this day. And so I must add a word of my own, not in the hope of contributing something to the issues which have been so exhaustively treated by the eminent men on both sides of the question, but in order that the true issue may be brought out in relief.

Reasoning by recurrence, whenever it is applied to finite sequences of numbers, is logically unassailable. In this *restricted* sense, the principle asserts that, if a proposition is of the hereditary type, then it is true or false of any term in the sequence if it is true or false of the first term in the sequence.

This *restricted principle* will suffice to create a finite, bounded arithmetic. For instance, we could terminate the natural sequence at the physiological or psychological limits of the counting process, say 1,000,000. In such an arithmetic addition and multiplication, when possible, would be *associative* and *commutative*; but the operations would not always be possible. Such expressions as $(500,000 + 500,001)$ or (1000×1001) would be meaningless, and it is obvious that the number of meaningless cases would far exceed those which have a meaning. This restriction on integers would cause a corresponding restriction on fractions; no decimal fraction could have more than 6 places, and the conversion of such a fraction as $1/3$ into a decimal fraction would have no meaning. Indefinite divisibility would have no more meaning than indefinite growth, and we would reach the indivisible by dividing any object into a million equal parts.

A similar situation would arise in geometry if instead of conceiving the plane as indefinitely extending in all direction we should limit ourselves to a *bounded region* of the plane, say a circle. In such a bounded geometry the intersection of two lines would be a matter of probability; two lines taken at random would not determine an angle; and three lines taken at random would not determine a triangle.

Yet, not only would such a bounded arithmetic and such a bounded geometry be logically impregnable, but strange though it may seem at first, they would be closer to the reality of our senses than are the unbounded varieties which are the heritage of the human race.

The restricted principle of mathematical induction involves a finite chain of syllogisms, each consistent in itself: for this reason the principle is a consequence of classical logic.

But the method used in the demonstrations of arithmetic, the *general* principle of complete induction, goes far beyond the confines imposed by the restricted principle. It is not content to say that a proposition true for the number 1 is true for all numbers, provided that if true for any number it is true for the successor of this number. *It tacitly asserts that any number has a successor.*

This assertion is not a logical necessity, for it is not a consequence of the laws of classical logic. This assertion does *not* impose itself as the only one conceivable, for its opposite, the postulation of a finite series of numbers, leads to a bounded arithmetic which is just as tenable. This assertion is *not* derived from the immediate experience of our senses, for all our experience proclaims its falsity. And finally this assertion is *not* a consequence of the historical development of the experimental sciences, for all the latest evidence points to a bounded universe, and in the light of the latest discoveries in the structure of the atom, the infinite divisibility of matter must be declared a myth.

And yet the concept of infinity, though not imposed upon us either by logic or by experience, is a *mathematical necessity.* What is, then, behind this power of the mind to conceive the indefinite repetition of an act when this act is once possible? To this question I shall return again and again throughout this study.

CHAPTER 5

Symbols

"One cannot escape the feeling that these mathematical formulae have an independent existence and an intelligence of their own, that they are wiser than we are, wiser even than their discoverers, that we get more out of them than was originally put into them."

—Heinrich Hertz

Algebra in the broad sense in which the term is used today, deals with operations upon symbolic forms. In this capacity it not only permeates all of mathematics, but encroaches upon the domain of formal logic and even of metaphysics. Furthermore, when so construed, algebra is as old as man's faculty to deal with general propositions; as old as his ability to discriminate between *some* and *any*.

Here, however, we are interested in algebra in a much more restricted sense, that part of general algebra which is very properly called the *theory of equations*. It is in this narrower sense that the term algebra was used at the outset. The word is of Arabic origin. "Al" is the Arabic article *the,* and *gebar* is the verb *to set,* to restitute. To this day the word "Algebrista" is used in Spain to designate a bonesetter, a sort of chiropractor.

It is fitting that the word algebra should be the adaptation of the title of a book written by Mohammed ben Musa Al Kworesmi, the same Al Kworesmi who, as we saw, contributed so much to the development of positional numeration. The full title of the book is "Algebar wal Muquabalah," an exact translation of which would read "On Restitution and Adjustment." Ben Musa used restitution in the same sense in which we today use *transposition,* i.e., the shifting of the terms of an equation from one side to the other, as for instance, the passing from $3x + 7 = 25$ to $3x = 25 - 7$.

Traces of a primitive algebra are found on the clay tablets of the Sumerians, and it probably reached quite a high degree of development among the ancient Egyptians. Indeed, the papyrus Rhind, of not later date than the eighteenth century B.C. deals with problems in *distribution* of food and other supplies, problems which lead to simple equations. The unknown in these equations is designated by *hau,* a heap; addition and subtraction by the legs of a man walking either towards the symbol of the operand or away from it. The papyrus is signed by one Ahmes. Ahmes, however, judging by the many gross errors in the text, was a mere scribe who understood little of what he was copying. So it is conjectured that the state of ancient Egyptian knowledge was higher than this papyrus would lead us to believe. Be this as it may, there is no doubt that Egyptian algebra antedates the papyrus by many centuries.

It is generally true that algebra in its development in individual countries passed successively through three stages: the *rhetorical,* the *syncopated,* and the *symbolic.* Rhetorical algebra is characterized by the complete absence of any symbols, except, of course, that the words themselves are being used in their symbolic sense. To this day rhetorical algebra is used in such statement as "the sum is independent of the order of the terms," which in symbols would be designated by $a + b = b + a$.

Operations / Modern Symbols	Addition	Subtraction	Multiplication	Division	Exponents	Equality	Unknown
Modern Symbols	$+$	$-$	$x \cdot a \cdot b$	$: \div \frac{a}{b}$	a^2, a^3	$=$	x, y, z
Source — Century							
Egyptian (17th B.C.)				$\frac{1}{3}$			
Diophantos of Alexandria		⋀		$\frac{1}{3} = \Upsilon^\theta$			ς
Hindu (11th)	The Sanskrit sound ya	A dot above the number			$x^2 = \square$	− − −	
Italian (16th)	$\tilde{P}$	$\tilde{m}$					
German (16th)	$+$	$-$					
Stevin (Belgium) (16th)	$+$	$-$		$\frac{3}{4}$	$x = ②$, $x = ③$	teta egale	⓪
Recorde (England) (16th)	$+$	$-$		$\frac{3}{4}$		$=$	
Vieta (France) (17th)	$+$	$-$	in		D^2 in quad. ratium	Aequabantur	$A, E, O,$
Oughtred (England) (17th)	$+$	$-$	$\times$	$\frac{3}{4}$	$x^4 = ▣$		
Harriot (England) (17th)	$+$	$-$			$a^2 = a^2$, $a^3 = a^3$	$=$	$a, b, d,$
Descartes (France) (17th)	$+$	$-$		$\frac{3}{4}$	$x^2 = x^2$ or xx	⊗	x, y, z
Leibnitz (Germany) (18th)	$+$	$-$		$\frac{a}{b}$	$a^3 = ③a$	$=$	any letter

Evolution of Symbols

Syncopated algebra, of which the Egyptian is a typical example, is a further development of rhetorical. Certain words of frequent use are gradually abbreviated. Eventually these abbreviations become contracted to the point where their origin has been forgotten, so that the symbols have no obvious connection with the operation which they represent. The *syncopation has become a symbol.*

The history of the symbols + and − may illustrate the point. In medieval Europe the latter was long denoted by the full word *minus,* then by the first letter *m* duly superscribed. Eventually the letter itself was dropped, leaving the superscript only. The sign *plus* passed through a similar metamorphosis. The reader is referred to the accompanying table for a chronological history of the standard symbols.

Greek algebra before Diophantus was essentially rhetorical. Various explanations were offered as to why the Greeks were so inept in creating a symbolism. One of the most current theories is that the letters of the Greek alphabet stood for numerals and that the use of the same letters to designate general quantities would have obviously caused confusion. It is pointed out that Diophantus took advantage of the fact that in Greek the sound ς (sigma) has two written forms: σ and ς: σ designated 60, but the end sigma, ς, had no numerical value, and it is for this reason that Diophantus chose it to symbolize the unknown.

The truth of the matter is that the Diophantine symbol for the unknown is more likely a syncopation of the first syllable of the Greek word *arithmos,* number, by which name he designated the unknown of a problem. Besides, the theory seems to disregard the fact that only the small letters of the Greek alphabet were used as numerals. The Greeks had at their disposal the capital letters which they could, and indeed did, use as symbols.

Yet those symbols were never used in an *operational* sense but merely as *labels,* to designate different points or elements of a geometrical configuration. Such descriptive symbols are used by us today in identifying various points of a geometrical figure, and it should be remembered that we inherited this custom from the Greeks.

No! Greek thought was essentially non-algebraic, because it was so concrete. The abstract operations of algebra, which deal with objects that have purposely been stripped of their physical content, could not occur to minds which were so intensely interested in the objects themselves. *The symbol is not a mere formality;* it is the very essence of algebra. Without the symbol the object is a human perception and reflects all the phases under which the human senses grasp it; replaced by a symbol the object becomes a complete abstraction, a mere *operand* subject to certain indicated operations.

Greek thought was just beginning to emerge from the plastic state, when the period of decadence set in. In these declining days of Hellenic culture two persons stand out. Both lived in the third century of our era, both hailed from Alexandria, both sowed the seeds of new theories, far too advanced to be absorbed by their contemporaries, but destined to grow into important sciences many centuries later. The "porisms" of Pappus anticipated *projective geometry,* the problems of Diophantus prepared the ground for the modern theory of equations.

Diophantus was the first Greek mathematician who frankly recognized fractions as numbers. He was also the first to handle in a systematic way not only simple equations, but quadratics

and equations of a higher order. In spite of his ineffective symbolism, in spite of the inelegance of his methods, he must be
regarded as the precursor of modern algebra.

But Diophantus was the last flicker of a dying candle. Over
the western world spread the long night of the Dark Ages. The
seeds of Hellenic culture were destined to sprout on alien soil.

The Hindus may have inherited some of the bare facts of Greek
science, but not the Greek critical acumen. Fools rush in where
angels fear to tread. The Hindus were not hampered by the
compunctions of rigor, they had no sophists to paralyze the
flight of their creative imagination. They played with number
and ratio, zero and infinity, as with so many words: the same
sunya, for instance, which stood for the void and eventually
became our zero, was also used to designate the unknown.

Yet the naïve formalism of the Hindus did more to develop
algebra than the critical rigor of the Greeks. It is true that theirs
was syncopated algebra, *par excellence.* The symbols were merely
the first syllables of the words designating the objects or operations; nevertheless they had symbols not only for the
fundamental operations and equality, but for negative numbers
as well. Moreover, they had developed all the rules for the transformation of simple and quadratic equations.

The types of problems they handled were simple enough,
and are really typical of that stage of algebra. We quote two from
Lilawati, a treatise on general theology written in the eighth century of our era:

"Out of a heap of pure lotus flowers one-third, one-fifth and
one-sixth were offered respectively to the gods Siva, Vishnu, and
the Sun; one-fourth was presented to Bhavani. The remaining six

flowers were given to the venerable preceptor. Tell me quickly the whole number of flowers." …

"A necklace was broken during an amorous struggle. One-third of the pearls fell to the ground, one-fifth stayed on the couch, one-sixth was found by the girl, and one-tenth recovered by her lover; six pearls remained on the string. Say of how many pearls the necklace was composed."

Hindu mathematics had little direct influence upon Europe. But there is little doubt that the Arabs got their arithmetic and algebra from the representatives of Brahmin knowledge who were so liberally entertained at the courts of the enlightened caliphs of the ninth and tenth centuries. Moslem civilization of that period was the blending of two cultures: the Oriental and the Hellenic. A great number of Sanskrit and Greek classics of literature, science and philosophy were translated into Arabic and avidly studied by the Arab savants. Many of these translations have been preserved and are now a fertile source of historical information. We must remember in this connection that the richest library of Hellenic antiquity, that of Alexandria, was twice pillaged or destroyed: first by Christian vandals in the fourth century, then by Moslem fanatics in the seventh. As a result of this destruction, a great number of ancient manuscripts disappeared, and would have been completely lost to posterity if it were not for their Arabic translations.

It has been often said that the historic destiny of the Arabs was to act as custodians of Hellenic culture during these transition ages. This they did exceedingly well. But they also enriched the treasure by brilliant contributions of their own. I may mention among the numerous first-rate mathematicians of the period the name of one man whose fame is familiar to every cultured person: Omar Khayyám. The author of the *Rubaiyat* was official astronomer of the court of the caliph. Though the *Rubaiyat* was written in Persian, Omar wrote an Arabic algebra in which he took full advantage of his knowledge of Greek geometry and Hindu algebra to solve cubic and quartic equations. Indeed, he can be considered as the originator of graphical methods. Furthermore, there are indications that he anticipated Newton in the discovery of the *binomial* formulas.

And yet for all this, the Arabs have not advanced one iota in symbolic notation. It is one of the strangest phenomena in the history of mathematics that the Arabs, in adopting Hindu algebra, did not retain their quaint syncopated symbolism. Quite the contrary; they dropped back to the rhetorical algebra of the Greeks and for a time even went so far as to eliminate numeral symbols from their treatises on algebra, preferring to write numbers out in full. Was it that the Arabs pushed their claim of being the intellectual heirs of the Hellenes to the point of refusing to acknowledge the debt they owed to the Brahmins?

While Moslem culture was approaching its highest point, Europe was still in deep slumber. A magnificent pen-picture of these dark ages and the centuries of transition which followed is given by the great mathematician Jacobi in his address on Descartes:

"History knew a midnight, which we may estimate at about the year 1000 A.D., when the human race had lost the arts and sciences even to the memory. The last twilight of paganism was gone, and yet the new day had not begun. Whatever was left of culture in the world was found only with the Saracens, and a Pope eager to learn studied in disguise at their universities, and so became the wonder of the West. At last Christendom, tired of praying to the dead bones of the martyrs, flocked to the tomb of the Saviour Himself, only to find for a second time that the grave was empty and that Christ had risen from the dead. Then mankind too rose from the dead. It returned to the activities and the business of life; there was a feverish revival in the arts and in the crafts. The cities flourished, a new citizenry was founded. Cimabue rediscovered the extinct art of painting; Dante, that of poetry. Then it was, also, that great courageous spirits like Abelard and Saint Thomas Aquinas dared to introduce into Catholicism the concepts of Aristotelean logic, and thus founded scholastic philosophy. But when the Church took the sciences under her wing, she demanded that the forms in which they moved be subjected to the same unconditioned faith in authority as were her own laws. And so it happened that scholasticism, far from freeing the human spirit, enchained it for many centuries to come, until the very possibility of free scientific research came to be doubted. At last, however, here too daylight broke, and mankind, reassured, determined to take advantage of its gifts and to create a knowledge of nature based on independent thought. The dawn of this day in history is known as the Renaissance or the Revival of Learning."

Now, the acquisition of culture was certainly not a part of the Crusader's program. Yet, this is exactly what the Crusades accomplished. For three centuries the Christian powers tried by sword to impose their "culture" upon Moslem. But the net result was that the superior culture of the Arabs slowly yet surely penetrated into Europe. The Arabs of Spain and the Arabs of the Levant were largely responsible for the revival of European learning.

The revival of learning began in Italy. The first notable work in mathematics was that done by Fibonacci, a man of extraordinary ability, whose insight and foresight were far above the thirteenth century in which he lived. A merchant by vocation, he had traveled considerably in the Near East and had absorbed the Arabic knowledge of the period; but he was also conversant with Greek mathematical literature. His contribution to arithmetic algebra, and geometry formed the rich source of Italian mathematics for the next three centuries. Of this, however, I shall speak in the next chapter.

The turning-point in the history of algebra was an essay written late in the sixteenth century by a Frenchman, Viète, who wrote under the Latin name Franciscus Vieta. His great achievement appears simple enough to us today. It is summed up in the following passage from this work:

> "In this we are aided by an artifice which permits us to distinguish given magnitudes from those which are unknown or sought, and this by means of a symbolism which is permanent in nature and clear to understand,—for instance, by denoting the unknown magnitudes by A or any other vowels, while the given magnitudes are designated by B, C, G or other consonants."

This vowel-consonant notation had a short existence. Within a half a century of Vieta's death appeared Descartes's *Géometrie,* in which the first letters of the alphabet were used for given quantities, the last for those unknown. The Cartesian notation not only displaced the Vietan, but has survived to this day.

But while few of Vieta's proposals were carried out in letter, they certainly were adopted in spirit. The systematic use of letters for undetermined but constant magnitudes, the "Logistica Speciosa" as he called it, which has played such a dominant rôle in the development of mathematics, was the great achievement of Vieta.

The lay mind may find it difficult to estimate the achievement of Vieta at its true value. Is not the literal notation a mere formality after all, a convenient shorthand at best? There is, no doubt, economy in writing

$$(a + b)^2 = a^2 + 2ab + b^2,$$

but does it really convey more to the mind than the verbal form of the same identity: the square of the sum of two numbers equals the sum of the squares of the numbers, augmented by twice their product?

Again, the literal notation had the fate of all very successful innovations. The universal use of these makes it difficult to conceive of a time when inferior methods were in vogue. Today formulæ in which letters represent general magnitudes are almost as familiar as common script, and our ability to handle symbols is regarded by many almost as a natural endowment of any intelligent man; but it is natural only because it has become a fixed habit of our minds. In the days of Vieta this notation constituted a radical departure from the traditions of ages. And really, how can we call natural a device which completely escaped the great Diophantus and his acute Arabian successors, while the ingenious Fibonacci was on the very brink of discovering it, but passed it up!

There is a striking analogy between the history of algebra and that of arithmetic. There, we saw, humanity struggled for thousands of years with an inadequate numeration for lack of symbol for naught. Here, the absence of a general notation reduced algebra to a collection of haphazard rules for the solution of numerical equations. Just as the discovery of zero created the arithmetic of today, so did the literal notation usher in a new era in the history of algebra.

Wherein lies the power of this symbolism?

First of all, the letter liberated algebra from the slavery of the word. And by this I do not mean merely that without the literal notation any general statement would become a mere flow of verbiage, subject to all the ambiguities and misinterpretations of human speech. This is important enough; but what is still more important is that the letter is free from the taboos which have attached to words through centuries of use. The *arithmos* of Diophantus, the *res* of Fibonacci, were preconceived notions: they meant a whole number, an integer. But the *A* of Vieta or our present x has an existence independent of the concrete object which it is assumed to represent. The symbol has a meaning which transcends the object symbolized: that is why *it is not a mere formality.*

In the second place, the letter is susceptible of operations which enables one to transform literal expressions and thus to paraphrase any statement into a number of equivalent forms. It is this power of transformation that *lifts algebra above the level of a convenient shorthand.*

Before the introduction of literal notation, it was possible to speak of individual expressions only; each expression, such as $2x + 3$; $3x - 5$; $x^2 + 4x + 7$; $3x^2 - 4x + 5$, had an individuality all its own and had to be handled on its own merits. The literal notation made it possible to pass from the individual to the collective, from the "some" to the "any" and the "all." The linear form $ax + b$, the quadratic form $ax^2 + bx + c$, each of these forms is regarded now as a single species. It is this that made possible the general theory of functions, which is the basis of all applied mathematics.

But the most important contribution of the *logistica speciosa,* and the one that concerns us most in this study, is the rôle it played in the formation of the generalized number concept.

As long as one deals with numerical equations, such as

(I) $x + 4 = 6$ (II) $x + 6 = 4$

$2x = 8$ $2x = 5$

$x^2 = 9$ $x^2 = 7,$

one can content himself (as most medieval algebraists did) with the statement that the first group of equations is possible, while the second is impossible.

But when one considers the literal equations of the same types:

$$x + b = a$$
$$bx = a$$
$$x^n = a$$

the very indeterminateness of the data compels one to give an *indicated* or *symbolic* solution to the problem:

$$x = a - b$$
$$x = a / b$$
$$x = \sqrt[n]{a}.$$

In vain, after this, will one stipulate that the expression $a - b$ has a meaning only if a is greater than b, that a/b is meaningless when a is not a multiple of b, and that $\sqrt[n]{a}$ is not a number unless a is a perfect nth power. The very act of writing down the *meaningless* has given it a meaning; and it is not easy to deny the existence of something that has received a name.

Moreover, with the reservation that $a > b$, that a is a multiple of b, that a is a perfect nth power, rules are devised for operating on such symbols as $a - b$; a/b; $\sqrt[n]{a}$. But sooner or later the very fact that there is nothing on the face of these symbols to indicate whether a legitimate or an illegitimate case is before us, will suggest that there is no contradiction involved in operating on these symbolic beings *as if they were bona fide numbers*. And from this

there is but one step to recognizing these symbolic beings as
numbers in extenso.

Such in its broad outlines is the story of early algebra, or rather
that phase of it which led up to the generalized number concept.
We shall now have to abandon the historical route, for two reasons. First of all, so rapid was the development of mathematics
after the days of Vieta that the systematic exposition of it would
lead us far beyond the scope of this book. Moreover, the foundation of the science of number was influenced but little by this
development, as long as progress was confined to technique
only.

What distinguishes modern arithmetic from that of the pre-
Vieta period is the changed attitude towards the "impossible."
Up to the seventeenth century the algebraists invested this term
with an absolute sense. Committed to natural numbers as the
exclusive field for all arithmetic operations, they regarded possibility, or restricted possibility, as an intrinsic property of these
operations.

Thus, the *direct* operations of arithmetic—addition $(a + b)$,
multiplication (ab), potentiation (a^b)—were *omnipossible*;
whereas the inverse operations—subtraction $(a - b)$, division
(a/b), extraction of roots $\sqrt[b]{a}$,—were possible only under restricted conditions. The pre-Vieta algebraists were satisfied with stating
these facts, but were incapable of a closer analysis of the problem.

Today we know that possibility and impossibility have each
only a relative meaning; that neither is an intrinsic property of
the operation but merely a *restriction which human tradition has
imposed on the field of the operand.* Remove the barrier, extend
the field, and the impossible becomes possible.

The direct operations of arithmetic are omnipossible because they are but a succession of *iterations*, a step-by-step penetration into the sequence of natural numbers, which is assumed *a priori* unlimited. Drop this assumption, restrict the field of the operand to a finite collection (say to the first 1000 numbers), and operations such as $925 + 125$, or 67×15 become impossible and the corresponding expressions meaningless.

Or, let us assume that the field is restricted to odd numbers only. Multiplication is still omnipossible, for the product of any two odd numbers is odd. However, in such a restricted field addition is an altogether impossible operation, because the sum of any two odd numbers is never an odd number.

Yet, again, if the field were restricted to *prime* numbers, multiplication would be impossible, for the simple reason that the product of two primes is never a prime; while addition would be possible only in such rare cases as when one of the two terms is 2, the other being the smaller of a couple of twin-primes, like $2 + 11 = 13$.

Other examples could be adduced, but even these few will suffice to bring out the *relative* nature of the words possible, impossible, and meaningless. And once this relativity is recognized, it is natural to inquire whether through a proper extension of the restricted field the inverse operations of arithmetic may not be rendered as omnipossible as the direct are.

To accomplish this with respect to subtraction it is sufficient to *adjoin to the sequence of natural numbers zero and the negative integers.* The field so created is called the general *integer field.*

Similarly, the adjunction of positive and negative *fractions* to this integer field will render division omnipossible.

The numbers thus created—the integers, and the fractions, positive and negative, and the number zero—constitute the *rational domain*. It supersedes the natural domain of integer arithmetic. The four fundamental operations, which heretofore applied to integers only, are now by *analogy* extended to these generalized numbers.

All this can be accomplished without a contradiction. And, what is more, with a single reservation which we shall take up presently, *the sum, the difference, the product, and the quotient of any two rational numbers are themselves rational numbers.* This very important fact is often paraphrased into the statement: the rational domain is *closed* with respect to the fundamental operations of arithmetic.

The single but very important reservation is that of division by zero. This is equivalent to the solution of the equation $x \cdot 0 = a$. If a is not zero the equation is *impossible*, because we were compelled, in defining the number zero, to admit the identity $a \cdot 0 = 0$. There exists therefore *no* rational number which satisfies the equation $x \cdot 0 = a$.

On the contrary, the equation $x \cdot 0 = 0$ is satisfied for any rational value of x. Consequently, x is here an *indeterminate* quantity. Unless the problem that led to such equations provides some further information, we must regard $0/0$ as the symbol of *any* rational number, and $a/0$ as the symbol of *no* rational number.

Elaborate though these considerations may seem, in symbols they reduce to the following concise statement: if a, b, and c are any rational numbers, and a is not 0, then there always exists a rational number x, and *only one*, which will satisfy the equation

$$ax + b = c$$

This equation is called *linear,* and it is the simplest type in a great variety of equations. Next to linear come quadratic, then cubic, quartic, quintic, and generally *algebraic* equations of any degree, the degree *n* meaning the highest power of the unknown *x* in

$$ax^n + bx^{n-1} + cx^{n-2} + \ldots + px + q = 0$$

But even these do not exhaust the infinite variety of equations; *exponential, trigonometric, logarithmic, circular, elliptic,* etc., constitute a still vaster variety, usually classified under the all-embracing term *transcendental.*

Is the rational domain adequate to handle this infinite variety? We shall see in the next chapter that this is emphatically not the case. We must anticipate an extension of the number domain to greater and greater complexity. But this extension is not arbitrary; there is concealed in the very mechanism of the generalizing scheme a guiding and unifying idea.

This idea is sometimes called the *principle of permanence.* It was first explicitly formulated by the German mathematician, Hermann Hanckel, in 1867, but the germ of the idea was already contained in the writings of Sir William Rowan Hamilton, one of the most original and fruitful minds of the nineteenth century.

I shall formulate this principle as a definition:

A collection of symbols infinite in number shall be called a *number field,* and each individual element in it a *number,*

First: If among the elements of the collection we can identify the sequence of *natural numbers.*

Second: If we can establish criteria of rank which will permit us to tell of any two elements whether they are equal, or if not equal, which is greater, these criteria reducing to the natural criteria when the two elements are *natural* numbers.

Third: If for any two elements of the collection we can devise a scheme of *addition* and *multiplication* which will have the commutative, associative, and distributive properties of the natural operations bearing these names, and which will reduce to these natural operations when the two elements are natural numbers.

These very general considerations leave the question open as to how the principle of permanence operates in special cases. Hamilton pointed the way by a method which he called *algebraic pairing*. We shall illustrate this on the rational numbers.

If *a* is a multiple of *b*, then the symbol *a/b* indicates the operation of division of *a* by *b*. Thus 9/3 = 3 means that the quotient of the indicated division is 3. Now, given two such indicated operations, is there a way of determining whether the results are equal, greater, or less, without actually performing the operations? Yes; we have the following

$$
\text{Criteria of Rank:} \begin{cases} \dfrac{a}{b} = \dfrac{c}{d} & \text{if} \quad ad = bc \\[2ex] \dfrac{a}{b} > \dfrac{c}{d} & \text{``} \quad ad > bc \\[2ex] \dfrac{a}{b} < \dfrac{c}{d} & \text{``} \quad ad < bc \end{cases}
$$

And we can go even go further that that: without performing the indicated operations we can devise rules for manipulating on these indicated quantities:

$$
\text{Addition:} \quad \frac{a}{b} + \frac{c}{d} = \frac{ad + bc}{bd}
$$

$$
\text{Multiplication:} \quad \frac{a}{b} \cdot \frac{c}{d} = \frac{ac}{bd}
$$

Now let us *not* stipulate any more that *a* be a multiple of *b*. Let us consider *a/b* as the symbol of a new field of mathematical beings. These symbolic beings depend on two integers *a* and *b* written in proper order. We shall impose on this collection of *couples* the criteria of rank mentioned above: i.e., we shall claim that, for instance:

$$\frac{20}{15} = \frac{16}{12} \quad \text{because} \quad 20 \times 12 = 15 \times 16$$

$$\frac{4}{3} > \frac{5}{4} \quad \text{because} \quad 4 \times 4 > 3 \times 5$$

We shall *define* the operations on these couples in accordance with the rules which, as we have shown above, are true for the case when *a* is a multiple of *b*, and *c* is a multiple of *d*; i.e., we shall say for instance:

$$\frac{2}{3} + \frac{4}{5} = \frac{(2 \times 5) + (3 \times 4)}{5 \times 3} = \frac{22}{15}$$

$$\frac{2}{3} \times \frac{4}{5} = \frac{2 \times 4}{3 \times 5} = \frac{8}{15}$$

We have now satisfied all the stipulations of the principle of permanence:

1. The new field contains the natural numbers as a sub-field, because we can write any natural number in the form of a couple:

$$\frac{1}{1}, \frac{2}{1}, \frac{3}{1}, \frac{4}{1}, \dots$$

2. The new field possesses criteria of rank which reduce to the natural criteria when *a/b* and *c/d* are natural numbers.

3. The new field has been provided with two operations which have all the properties of addition and multiplication, to which they reduce when a/b and c/d are natural numbers.

And so these new beings satisfy all the stipulations of the principle. They have proved their right to be adjoined to the natural numbers, their right to be invested with the dignity of the name *number*. They are therewith admitted, and the field of numbers comprising both old and new is christened the *rational domain of numbers.*

It would seem at first glance that the principle of permanence leaves such a latitude in the choice of operations as to make the general number it postulates too general to be of much practical value. However, the stipulations that the natural sequence should be a part of the field, and that the fundamental operations should be commutative, associative, and distributive (as the natural operations are), impose restrictions which, as we shall see, only very special fields can meet.

The position of arithmetic, as formulated in the principle of permanence, can be compared to the policy of a state bent on expansion, but desirous to perpetuate the fundamental laws on which it grew strong. These two different objectives—expansion on the one hand, preservation of uniformity on the other—will naturally influence the rules for admission of new states to the Union.

Thus, the first point in the principle of permanence corresponds to the pronouncement that the nucleus state shall *set the tone* of the Union. Next, the original state being an oligarchy in which every citizen has a rank, it imposes this requirement on the new states. This requirement corresponds to the second point of the principle of permanence.

Finally, it stipulates that the laws of commingling between the citizens of each individual state admitted to the Union shall be of a type which will permit unimpeded relations between citizens of that state and those of the nucleus state.

Of course I do not want the reader to take this analogy literally. It is suggested in the hope that it may invoke mental associations from a more familiar field, so that the principle of permanence may lose its seeming artificiality.

The considerations, which led up to the construction of the rational domain, were the first steps in a historical process called the *arithmetization of mathematics.* This movement, which began with Weierstrass in the sixties of the last century, had for its object the separation of purely mathematical concepts, such as *number* and *correspondence* and *aggregate,* from intuitional ideas, which mathematics had acquired from long association with geometry and mechanics.

These latter, in the opinion of the *formalists,* are so firmly entrenched in mathematical thought that in spite of the most careful circumspection in the choice of words, the meaning concealed behind these words may influence our reasoning. For the trouble with human words is that they *possess content,* whereas the purpose of mathematics is to construct pure forms of thought.

But how can we avoid the use of human language? The answer is found in the word *symbol.* Only by using a symbolic language not yet usurped by those vague ideas of *space, time, continuity* which have their origin in intuition and tend to obscure pure reason—only thus may we hope to build mathematics on the solid foundation of logic.

Such is the platform of this school, a school which was founded by the Italian Peano and whose most modern representatives are Bertrand Russell and A. N. Whitehead. In the fundamental work of the latter men, the *Principia Mathematica*, they have endeavored to reconstruct the whole foundation of modern mathematics, starting with clear-cut, fundamental assumptions and proceeding on principles of strict logic. The use of a precise symbolism should leave no room for those ambiguities which are inseparable from human language.

The *Principia* will long remain a monument of hard labor and excellent intentions. Have its authors succeeded in erecting a structure reared on pure reason and untainted by human intuition? I am incompetent to answer this question, as I have never met a mathematician who has read all three of its volumes. The story current in mathematical circles is that there are but two people who have ever read the *Principia* from cover to cover. Whether the authors themselves were included in this estimate I have not been able to ascertain.

I confess that I am out of sympathy with the extreme formalism of the Peano-Russell school, that I have never acquired the taste for their methods of symbolic logic, that my repeated efforts to master their involved symbolism have invariably resulted in helpless confusion and despair. This personal ineptitude has undoubtedly colored my opinion—a powerful reason why I should not air my prejudices here.

Yet I am certain that these prejudices have not caused me to underestimate the rôle of mathematical symbolism. To me, the tremendous importance of this symbolism lies not in these sterile attempts to banish intuition from the realm of human thought, but in its unlimited power to *aid intuition* in creating new forms of thought.

To recognize this, it is not necessary to master the intricate technical symbolism of modern mathematics. It is sufficient to contemplate the more simple, yet much more subtle, symbolism of language. For, in so far as our language is capable of precise statements, it is but a system of symbols, a rhetorical algebra *par excellence.* Nouns and phrases are but symbols of classes of objects, verbs symbolize relations, and sentences are but propositions connecting these classes. Yet, while the word is the abstract symbol of a class, it has also the capacity to *evoke an image,* a concrete picture of some representative element of the class. It is in this dual function of our language that we should seek the germs of the conflict which later arises between logic and intuition.

And what is true of words generally is particularly true of those words which represent *natural* numbers. Because they have the power to evoke in our mind images of concrete collections, they appear to us so rooted in firm reality as to be endowed with an *absolute* nature. Yet in the sense in which they are used in arithmetic, they are but a set of abstract symbols subject to a system of operational rules.

Once we recognize this symbolic nature of the natural number, it loses its absolute character. Its intrinsic kinship with the wider domain of which it is the nucleus becomes evident. At the same time the successive extensions of the number concept become steps in an inevitable process of natural evolution, instead of the artificial and arbitrary *legerdemain* which they seem at first.

CHAPTER 6

The Unutterable

"God created the integers, the rest is the work of man."

—Leopold Kronecker

Number ruled the universe of the Pythagoreans.

Not number in the modern sense of the word: it was the natural number, the integer, that reigned supreme. But neither was the universe of the Pythagoreans our universe, a universe which transcends the immediate sense-perception, which manifests itself so richly, even if mysteriously, in the numerous inventions which make up the essential part of our daily life: the universe of the Greeks was limited to things more immediately accessible to the senses.

In the harmonies of sound the Pythagoreans saw a confirmation of their number philosophy. The harmony of sight and touch found supreme expression in the perfect figures of geometry: the circle and the sphere, the regular polygons and the perfect solids, such were the elements used by the Great Architect in building the world. Here too, it was confidently expected, number would be found to reign supreme.

"The point is unity in position" was the basis of Pythagorean geometry. Behind this flowery verbiage we detect the naïve idea of the line as made up of a succession of atoms just as a necklace is made up of beads. The atoms may be ever so small, yet, being all homogeneous in substance and equal in size, they may be taken as the ultimate unit of measure. Therefore, given any two segments, the ratio of their length was merely the ratio of the numbers of atoms in each.

The same, of course, was true of the sides of any triangle, particularly the right triangle. From Egypt the Pythagoreans imported the "golden" triangle, the sides of which were in the ratio 3:4:5. Soon other "Pythagorean" triangles, such as 5:12:13 and 8:15:17, were discovered. The conviction that all triangles were *rational* had evidence to feed on. That some of the triangles, most of them in fact, did not yield such perfect ratios was not at all surprising; for after all, the ratios may run into very large numbers, and the calculating technique of the Greeks was primitive enough.

So matters stood for a while.

The contemplation of such triangles led to a capital discovery, which to this day bears the name of Pythagoras and which is one of the basic theorems of classical geometry. It reads: *In any right triangle the sum of the squares built on the legs is equal to the square built on the hypothenuse.* According to legend, the theorem was discovered by Pythagoras himself, who was so overwhelmed by its elegance that he sacrificed an ox to the gods. How this legend can be reconciled with the pretty well established fact that the Pythagoreans were strict vegetarians is left to the reader.

It may be doubted that Pythagoras derived his theorem through deductive reasoning. Most likely it was an empirical product. That he possessed a rigorous proof of the proposition is just as unlikely. But there can be little doubt that he and his disciples attached the greatest importance to it; for *therein they saw the inherent union between geometry and arithmetic,* a new confirmation of their dictum: "Number rules the universe."

But the triumph was short-lived. Indeed, one of the immediate consequences of the theorem was another discovery: *the diagonal of the square is incommensurable with its side.* Who it was that first established this, and how it was done, will probably remain a mystery forever. Euclid's beautiful proof, which is given below, is obviously the development of a cruder method. But whoever discovered it, there is little doubt that it caused great consternation in the ranks of the Pythagoreans. The very name given to these entities testifies to that. *Alogon,* the *unutterable,* these incommensurables were called, and the members of the order were sworn not to divulge their existence to outsiders. An unaccountable imperfection having been uncovered in the work of the Architect, it must needs be kept in strict concealment, lest His wrath at being exposed be visited upon man.

Says Proclos:

"It is told that those who first brought out the irrationals from concealment into the open perished in shipwreck, to a man. For the unutterable and the formless must needs be concealed. And those who uncovered and touched this image of life were instantly destroyed and shall remain forever exposed to the play of the eternal waves."

Less than a century passed, and the Pythagorean secret became the property of all thinking men. The unutterable had been spoken, the unthinkable clothed in words, the unrevealable presented to the eyes of the unitiated. Man had tasted of the forbidden fruit of knowledge and was condemned to be banished from the Pythagorean number paradise.

The advent of irrationals marks the decline of Pythagoreanism as a system of natural philosophy. That perfect concordance between things arithmetical and things geometrical, which the Pythagoreans preached, turned out to be a hoax: How can number dominate the universe when it fails to account for even the most immediate aspect of the universe, *geometry?*

So ended the first attempt to exhaust nature by number.

Like most classical demonstrations, Euclid's proof of the incommensurability of the diagonal of the square with its side is of the type *reductio ad absurdum.* It is geometrical only in appearance, for it is based on pure consideration of the theory of numbers. In modern phraseology, each side of the square being taken as one, and the diagonal denoted by x, Pythagoras' theorem reduces the problem to the solution of the quadratic equation:

(1) $x^2 = 1^2 + 1^2$, or $x^2 = 2$

If it is possible to satisfy this equation by a rational number p/q, then the diagonal and the side are commensurable. Let us assume that such is the case and that the fraction p/q is in its *lowest terms.* Then one of the two integers, either p or q, must be odd. I shall show that p cannot be odd. Indeed, substitute in equation (1)p/q for x; it then becomes

(2) $\dfrac{p^2}{q^2} = 2$ or $p^2 = 2q^2$

which shows that p^2, and therefore p, is even.

Now since p is even, we can set $p = 2r$, where r is another unknown integer. Substituting this in (2), we obtain

(3) $4r^2 = 2q^2$ or $q^2 = 2r^2$

which is of the same type as (2). But this means that the integer q is also even, which contradicts our assumption that p/q is in its lowest terms, which in turn shows the impossibility of satisfying the equation $x^2 = 2$ by a rational number.

The argument is perfectly general. Slightly modified it applies to the equation

$$x^2 = 3, x^2 = 5, x^2 = 6,$$
$$x^3 = 2, x^3 = 3, x^3 = 4$$

and more generally to the equation

$$x^n = a$$

When a is not the perfect nth power of some rational number, the equation $x^n = a$ has no rational solutions.

We find in the writings of the minor Greek geometers, such as Hero of Alexandria and Theon of Smyrna, approximate values for the irrational numbers $\sqrt{2}, \sqrt{3}, \sqrt{5}$ etc. No mention is made as to the method by which these values were obtained. Since in most of these cases the approximations were excellent, historians of mathematics have let their imagination play freely in reconstructing the unknown methods. There are quite a number of such theories. Some credit Greek mathematicians with the knowledge of infinite series; others, of continued fractions. I venture a theory of my own, which, while just as speculative, has at least the merit of not assuming that the Greeks were versed in modern methods.

My theory is this: Euclid's proof of the irrationality of $\sqrt{2}$ was, for the *average* Greek mathematician, too exotic to be convincing. There might have been among the Pythagoreans "die-hards" who had not given up the hope of finding a rational value for $\sqrt{2}, \sqrt{3}$, etc. The search for such rational values proceeded along most natural lines. The number 2, for instance, could be presented in an infinite number of ways as a fraction whose denominator was a perfect square:

$$\frac{2}{1} = \frac{8}{4} = \frac{18}{9} = \frac{32}{16} = \frac{50}{25} = \frac{72}{36} = \frac{128}{64} = \frac{200}{100} = \ldots$$

If $\sqrt{2}$ were a rational number, then by "going out" far enough a fraction would eventually be found whose numerator is also a perfect square. In this, of course, they failed, but as a by-product an excellent approximation was found. Indeed

$$\frac{288}{144} = 2,$$

while

$$\frac{289}{144} = \left(\frac{17}{12}\right)^2.$$

This gives for $\sqrt{2}$ the Theon approximation, $1\frac{5}{12}$, which differs from the true value by less than $\frac{1}{7}$ of 1%.

I offer this theory for what it is worth.

There is a great variety of problems in geometry, some of the most simple type, which admit of no *numerical* solution, at least not so long as we confine ourselves to the rational domain of numbers. Take the case of a diagonal of a square of side one as an instance. A child who had learned the elementary constructions by ruler and compass could determine the diagonal *geometrically.* The same is true of other problems that lead to quadratic, cubic, or higher equations, or even transcendental equations. Yet these problems escape completely the attack by *rational arithmetic.*

On the other hand, the irrational quantities can be expressed through rational approximations to any desired degree of accuracy. The procedure we presented in the section just above is perfectly general. Procedures of a similar character are: the *algorithm* for the extraction of a square root, which we are taught in school; expansion into series; continued fractions; and numerous other devices, which can be drafted into service whenever we are confronted with a problem which admits of no rational solution. Such methods enable one to "trap" the irrational number between two sequences of rational numbers, of which the first is consistently "less" than the irrational, and the second consistently "greater." And, what is more, the interval between these rational approximations may be rendered as small as one desires.

Well, what further can be desired? The physicist, the engineer, the practical man generally are fully satisfied. What the physicist requires of his calculating methods is a degree of refinement which will permit him to take full advantage of the growing precision of his measuring devices. The fact that certain magnitudes, like $\sqrt{2}$, π, or e, are not expressible *mathematically* by means of rational numbers will not cause him to lose any sleep, as long as mathematics is furnishing him with rational approximations for such magnitudes to any accuracy he desires.

The position of the mathematician with respect to this problem is different, and for this reason: he views the rational domain of numbers as a *totality*, as an *aggregate*. He sees this aggregate extending from negative infinity, through zero, into positive infinity. This aggregate is *ordered*: give him any two rational numbers and he will tell you which is the greater. Between any two rational numbers he can insert a third, and this no matter how near the two numbers may be. He expresses this in his jargon by saying that the rational domain is *everywhere dense.* In short, he views the aggregate of rational numbers as a *compact, continuous mass, seemingly without gaps.*

To him there is a striking analogy between this and the set of points on a straight line. Here too the aggregate extends indefinitely in both directions. Here too of any two elements he can tell which is to the right. Here too he finds the property of compactness; for between any two points he can insert a third, and this no matter how close the points may be together. So complete does this analogy appear, that there should be a way of establishing *a correspondence between the rational domain of numbers on the one hand and the points on a line on the other hand.*

This correspondence is the basis of *analytic geometry,* and even those of my readers who have never studied it have a very good idea of what it is all about from the occasional graphs they have handled. So I shall remind them only that the scheme consists in defining on an indefinite straight line a positive and a negative *sense.* Such a line endowed with "sense" is called an *axis.* On the axis we select two points: O, the *origin,* which represents the number zero; and U, the *unit point,* which represents the number 1. The positive integers we obtain by laying off to the right a succession of intervals equal to OU in length, and the negatives by proceeding similarly to the left. By dividing the unit segment into any number of aliquot parts we are able to represent any positive or any negative fraction as well.

Thus any rational number may be represented by a point on the axis; and immediately the question arises whether the converse is also true, i.e., whether to any point on the axis will correspond a rational number. The answer is an emphatic *no*: for if we construct a square of a side equal to OU and carry a segment OD equal to the diagonal of this square, to the point D will correspond *no rational number.*

So the absence of the gaps was only a delusion! It is true that if *all* rational numbers were *mapped* on the axis we would obtain a compact set; yet these points would by no means fill the line: there would be left an infinite number of gaps which would admit of no such representation. And we shall see later in what sense we may claim that *the irrational gaps by far exceed the rational points.*

From the standpoint of the pure mathematician the fundamental fact is this: To any rational number corresponds a point on the axis, but the correspondence is not reciprocal. There are points on the axis to which no rational number can be assigned: these points are not only infinite in number, but infite in variety as well, each variety, such as $\sqrt{a}, \sqrt[3]{a}, \ldots \sqrt[n]{a}$, etc., comprising an infinite number of *irrational* points.

Thus we are again confronted with the task of extending the number concept. We must push the number domain far beyond the concept of rational number, for this latter will not suffice even for the solution of the most simple quadratic equation.

It is natural to invoke once more the principle of permanence which rendered such valuable service before. We create the symbol $\sqrt[n]{a}$. This symbol represents a rational number if the equation $x^n = a$ has a rational solution; i.e., when a is a perfect nth power of a rational number. Using this special case as a point of departure, we establish rules for operating on these symbols. These identities are then taken as defining relations of a new number field, the field of *elementary irrationals,* the *radicals,* symbolized by $\sqrt[n]{a}$. Thanks to the device of reduction to a common exponent the criteria of rank are easily established. If for instance we wish to compare $\sqrt{2}$ and $\sqrt[3]{3}$ we write

$$\sqrt{2} = \sqrt[6]{8}; \quad \sqrt[3]{3} = \sqrt[6]{9}; \quad \sqrt[6]{9} > \sqrt[6]{8}$$

and this leads to the inequality $\sqrt[3]{3} > \sqrt{2}$.

Multiplication and division are also easily defined by means of the same device. The product of any two *surds* bearing on rational numbers is itself an entity of the same type. For instance,

$$\sqrt{2} \cdot \sqrt[3]{3} = \sqrt[6]{8 \times 9} = \sqrt[6]{72}.$$

An insurmountable difficulty, however, is met in addition. Such an expression as $\sqrt{2} + \sqrt{3}$ cannot be expressed in the form $\sqrt[n]{a}$ where a is a rational number. *The sum of two elementary irrationals is generally not an elementary irrational.* The field of simple irrationals is "closed" with respect to multiplication and division, but it is "wide open" to addition and subtraction.

To erect a consistent system we would be compelled to extend at once our field from elementary to *compound* irrationals of the type $\sqrt[n]{a} + \sqrt[n]{b}$. But before proceeding with such an extension let us take a glimpse of what is ahead of us.

We must not forget that we set out to "solve" the most general equation,

$$ax^n + bx^{n-1} + cx^{n-2} + \dots + px + q = 0,$$

where n is any integer and the coefficients are any rational numbers. Of these very general equations we have tackled so far the very special case of a *binomial* equation,

$$ax^n + b = 0.$$

The case $n = 1$ led us to the rational domain. The general case brought in elementary irrationals. But what about the general equations? If it be reducible to the equation of the simple type $x^n = A$, then its formal solution would lead to elementary irrationals. The fundamental question is this: Can any algebraic equation be reduced to binomial equations? Or, in other words, *can the solution of the general algebraic equation be formally expressed by means of radicals?*

The history of this problem provides an excellent example of the value of inductive inference.

Special types of quadratic equations are found in Diophantus' *Arithmetica*. The theory was further developed by the Hindus, who have in connection with it established the rules for the operations on surds almost in the form in which we know them today. The task was completed by the Arabian mathematicians: the formal solution of the *general quadratic* equation $ax^2 + bx + c = 0$ was found to be one of the form $A + \sqrt{B}$, i.e., it was *expressible by means of rational numbers and quadratic surds.*

The Arabians next tackled the general cubic equation $ax^3 + bx^2 + cx + d = 0$ with questionable success. Omar Khayyám gave an ingenious geometrical solution, but his futile efforts to obtain a formal algebraic solution led him to allege that the cubic equation cannot be solved by radicals. The problem greatly fascinated the Italian mathematicians of the Renaissance period, and it was completely solved by them in the sixteenth century under circumstances of which I tell elsewhere. They found that *the general solution of the cubic equation can be expressed by means of cubic and quadratic surds.*

Almost simultaneously the Italian Ferrari reduced the quartic equation to the solution of auxiliary quadratics and cubics, thus proving that *the equation of the fourth degree is capable of a formal solution by means of radicals the exponent of which does not exceed four.*

The natural inference was that this is generally true: the equation of degree n should, *by precedent,* be capable of a formal solution by means of radicals, and probably radicals of an exponent not higher than n. Such, indeed, was the general conviction of most mathematicians of the eighteenth century, one of the most notable exceptions being Lagrange.

The problem was not solved until the first part of the nineteenth century. As it frequently happens in mathematics, the extreme difficulty of the problem required new methods, and the new methods turned out to be much more fruitful and far-reaching than what the problem called for. The fundamental contributions of Ruffini, Abel, and Galois not only completely solved the problem, but enriched mathematics with a new and fundamental concept: *the group.*

It is most remarkable that two men as different in character and outlook as Abel and Galois should have become interested in the same problem and should have attacked it by similar methods. Both approached the problem of the *quintic* equation in the conviction that a solution by radicals was possible; Abel at eighteen, Galois at sixteen. In fact, both thought for a while that they had discovered such a solution; both soon realized their error and attacked the problem by new methods.

In 1825 Abel proved conclusively that *the general equation of the fifth degree cannot be solved by means of radicals only.* He surmised that the same is true of all equations higher than the fifth. This was definitely proved by Galois. The question, "What must be the special nature of an equation that it may lead to a solution by radicals?" was fully answered in Galois' testament-memoir. It is this special problem that led Galois to establish a new theory of equations, which is usually called the Galois theory of groups. This, however, is beyond our scope.

To resume then, the direct application of the principle of permanence to the problem of irrationals fails for two reasons. First, the elementary irrationals, i.e., those of the form $\sqrt[n]{a}$, do not constitute a closed field. In the second place, the *compound* irrational forms are inadequate for the solution of the general equation of degree *higher than the fourth*.

To create a comprehensive theory it is necessary to consider the entire *algebraic* field, a field which would contain all *algebraic* numbers, i.e., the solution of all possible algebraic equations. Such a field would certainly embrace the rational domain. Furthermore it can be shown that the algebraic field is closed with respect not only to the first four fundamental operations but also to root extraction; i.e., the sum, the difference, the product, the quotient, the powers, and the surds of any two algebraic numbers are themselves algebraic numbers. And what is more, if we consider the most general equation,

$$ax^n + bx^{n-1} + \ldots + px + q = 0,$$

where n is an integer, but a, b, c ... p, q, are no longer restricted to being rational numbers, but can themselves be algebraic numbers of the most general type; if such an equation admits of a solution at all, the *solution will be an algebraic number.*

Yet, comprehensive though the theory of algebraic numbers may be, it has several grave defects. In the first place the symbolism which it requires is vague and unwieldy, for it involves all the coefficients of the equation. Furthermore the operations on such symbols are so complicated as to make even the most simple manipulations impracticable. Finally, there is the very serious difficulty that an algebraic equation above the linear has generally *more than one solution.*

The quadratic equation may have two, the equation of the nth degree may have as many as n distinct solutions. The ambiguity inherent in such a procedure is an insurmountable obstacle, particularly in application to problems where *single valuedness* is a prime consideration.

But long before the movement to generalize the number concept in this direction had gained momentum, an event occurred which eclipsed in importance even the discoveries of Abel and Galois. In 1844 the French mathematician Jacques Liouville, professor at the École Normale and founder of the important *Journal des Mathématiques,* read before the Paris Academy a note which was later published in his own magazine under the title: "On the very extensive class of quantities which are neither algebraic nor even reducible to algebraic irrationals." In this epoch-making memoir Liouville exhibited quantities which by their very nature cannot be the roots of any algebraic equations, and thus confirmed a suspicion which Legendre had uttered as far back as 1794.

Rich though the variety of algebraic numbers may seem, they are but a province in a vaster domain, a much more extensive domain, as was shown fifty years later by Georg Cantor, who, in giving a new proof of the Liouville theorem, has put the theory of *transcendentals,* as these non-algebraic numbers are called, on a solid foundation. Of this, however, later.

Stranger yet is the fact that these transcendentals are not just a weird product of a mathematical imagination, a dish to whet the mathematician's appetite for abstraction. The invention of the calculus brought in its wake a class of quantities which during the succeeding centuries began to play a predominant rôle in practically every problem of analysis: the *logarithms* and the *trigonometric ratios.* Today these quantities are in daily use in every engineering office in the world and constitute a most powerful tool of applied mathematics. In the fifty years that followed Liouville's announcement, it was definitely established that most of these quantities were transcendentals. To understand this better let us take a glimpse at the history of the number π.

> "Also he made a molten sea of ten cubits from brim to brim, round in compass, and five cubits the height thereof; and a line of thirty cubits did compass it round about." (*Chronicles* IV, 2.)

This description of the priests' bathing pool in Solomon's Temple seems to indicate that the ancient Jews held that π, the ratio of the circumference of a circle to its diameter, equals three. This value is 5% short of the actual. The Egyptians made a much closer estimate: we find in the papyrus Rhind (1700 B.C.) the value of π as equal to $3^{13}/81$ or 256/81 or $(16/9)^2$, which is less than 1% in excess.

It is only natural that this quantity should have been the subject of speculation by the Greek mathematicians from the earliest times. But on Greek soil the problem acquired a new character. It took its place among the famous problems of antiquity, wrapped in all the legendary splendor of Greek mythology.

These problems were three in number: *the doubling of a cube, the trisection of an angle, and the squaring of a circle.* The last is substantially equivalent to the determination of π, for the area of a circle of unit radius is equal to π square units, and if

the number π could be expressed rationally the whole question
would be reduced to the construction of a square of a given area. In
fact, if the Egyptian value were true, the area of the circle would be
the same as that of a square built on 8/9 of its diameter.

Around these three problems grew up most of Greek geometry.
In the effort to solve them the Greek geometers discovered the conic
sections and a number of higher curves. Probably, they did not
suspect that a solution such as they sought did not exist, and the difficult and recalcitrant nature of the problem only spurred their
efforts and drew into the arena of geometry their greatest minds
from Archimedes to Apollonius of Perga.

Yet the first two problems are algebraically equivalent to the
solution of the relatively simple cubic equations: $x^3 - 2 = 0$, and
$4x^3 - 3x - a = 0$, where a is a proper fraction. When we brand
these problems as impossible, do we use the word "impossible"
in the same sense as we did in arithmetic where the impossibility was tantamount to a restriction imposed on the field?

Yes! The impossibility of the classical problems was imposed
by a restriction which was so old as to be considered natural, so
natural, indeed, that it was rarely mentioned. When the Greek
spoke of a geometrical construction, he meant a construction by
straight-edge and *compass*. These were the instruments of the
gods; all other means were banned as unworthy of the speculation of the philosopher. For Greek philosophy, we must remember,
was essentially aristocratic. The methods of the artisan, ingenious and elegant though they might appear, were regarded as
vulgar and banal, and general contempt attached to all those
who used their knowledge for gainful ends. (There is the story of
the young nobleman who enrolled in the academy of Euclid.

After a few days, he was so struck with the abstract nature of the subject that he inquired of the master of what practical use his speculations were. Whereupon the master called a slave and commanded: "Give this youth a chalcus, so that he may derive gain from his knowledge.")

Now, those problems which are capable of a solution by straight-edge only, problems which today are called *linear,* lead, when couched in algebraic language, to *linear* equations. But those which in addition require the use of the compass are algebraically equivalent to the solution of an equation of the *second degree.* These facts, however, were not known before the seventeenth century of our era. In the meantime the two problems were repeatedly attacked by the brightest minds and by others not so bright. To this day there are professional "trisectors" whose greatest handicap lies in the fact that they have never learned that the problem was disposed of three hundred years ago.

Now the fact that a *straight-edge-compass* solution of a geometrical problem is either linear or quadratic does not mean that if a problem leads to a higher equation it is incapable of such a solution. To illustrate, consider the equation $x^4 - 3x^2 + 2 = 0$. The left part of it factors into $(x^2 - 1)(x^2 - 2)$, and consequently this *quartic* equation really breaks up into two quadratics. Whenever such a manipulation is feasible, i.e., whenever we can separate from an equation an expression of *rational coefficients* with a lower order, the equation is said to be *reducible.*

The trouble with the cubic equations to which the doubling of the cube and the trisection of the general angle lead is that they are *irreducible,* and this fact condemns the problems which are behind these equations as *insoluble by ruler and compass.*

We have here another confirmation of the relative nature of the term *impossible.* Impossibility is nearly always the result of a restriction, usually a restriction so sanctified by tradition that it seems imposed by nature itself. Remove the restriction and the impossibility will disappear. So it is here too. It is known today that by means of special linkages, that is, instruments consisting of a series of pivoted rigid members, it is possible to solve not only these two problems, but any problem leading to an *algebraic equation with rational coefficients.*

The problem of squaring the circle is different from the other two in that it escapes altogether algebraic formulation.

Attempts to solve this problem fill the annals of mathematics since the days of Pythagoras. Archimedes was the first to recognize that the difficulty lies in the definition. When we speak of the area of a rectangle or of a triangle we can define our terms with precision; the same is true of any polygonal figure. But what do we mean by the area inclosed within a curve? It is true that we can inscribe or circumscribe polygonal lines and speak of *upper* or *lower bounds* for such an area; but the area itself cannot be defined without bringing in *infinite processes* and *limits.*

We shall see later that it was on this problem that Archimedes tested the power of the so-called method of *exhaustion.* Here it is enough to mention that by means of a series of polygons, some inscribed, others circumscribed, to a circle, Archimedes showed that π was contained between $3^{1}/7$ and $3^{10}/71$.

In the eighteen hundred years which followed Archimedes, the problem made little progress. There were always, of course, plenty of *circle-squarers*, and various alleged solutions were published, some of which were curious enough. Also there were many approximations, among which the most interesting was $\sqrt{10}$, probably of Hindu origin. This value, which is very close to the

Egyptian, was very much in use throughout the Middle Ages. Many efforts to improve on Archimedes are also recorded; not the least notable is that of Vieta, who used a polygon of 393,216 sides to obtain π within 10 correct decimals.

The invention of infinite processes had for result such a great refinement in calculations that Vieta's value was soon put into the shade. Today over 700 correct decimals of the number π are known. As to the practical value of such calculations, we leave the word to the American astronomer Simon Newcomb:

> "Ten decimals are sufficient to give the circumference of the earth to the fraction of an inch, and thirty decimals would give the circumference of the whole visible universe to a quantity imperceptible to the most powerful microscope."

From the theoretical standpoint a possible justification may be that such labors bespeak the refinement of modern mathematical methods. Also there may be the forlorn hope that we may discover some sort of regularity in the succession of the decimals which would throw light on the nature of the number π.

At the end of the eighteenth century the problem enters an entirely new phase. Lambert showed that π is *not a rational* number, and Legendre established the fact that it cannot be the root of a quadratic equation with rational coefficients. This *definitely disposed of the problem of squaring the circle,* without, of course, dampening in the least the ardor of the circle-squarers. For it is characteristic of these people that their ignorance equals their capacity for self-deception.

There was still the possibility that π was an algebraic number. Should such be the case, the squaring of the circle, impossible by compass and straight-edge would, formally at least, be amenable to a link-work solution. This, while of no practical value, would give a fitting climax to two thousand years of futile effort.

This possibility was considerably impaired when in 1873 the French mathematician Charles Hermite proved that the *number e was a transcendental.* The intimate connection between the numbers e and π was well known, and this redoubled the efforts to prove that π also was a transcendental number. This was indeed achieved by the German Lindemann nine years later. Thus did modern analysis dispose of a problem which had taxed the ability of mathematicians since the days of Thales.

So ended the second attempt to exhaust nature by number.

The discovery of transcendentals, the establishment of the fact that they are far richer in extent and variety than the irrationals of algebra, that they comprise some of the most fundamental magnitudes of modern mathematics—all this showed definitely that the powerful machinery of algebra had failed just where the elementary tools of rational arithmetic had failed two thousand years earlier. Both failures were due to the same source: algebra, like rational arithmetic, dealt with *finite processes* only.

Now, as then, infinity was the rock which wrecked the hope to establish number on a firmer foundation. But to legalize infinite processes, to admit these weird irrational creatures on terms of equality with rational numbers, was just as abhorrent to the rigorists of the nineteenth century as it had been to those of classical Greece.

Loud among these rose the voice of Leopold Kronecker, the father of modern *intuitionism.* He rightly traced the trouble to the introduction of irrationals and proposed that they be banished from mathematics. Proclaiming the absolute nature of the integers he maintained that the natural domain, and the rational domain immediately reducible to it, were the only solid foundation on which mathematics could rest.

"God made the integer, the rest is the work of man," is the famous phrase by which he will be best known to posterity. This phrase reminds me of the story of the pious old dame who was heading a committee for the erection of a new church. The architect who submitted the plans found that the old lady took the business very seriously. Most vehement was her protest against the stained glass for which his specifications called. Finally in despair he asked her on what ground she objected to stained glass. "I want my glass the way the Lord made it!" was her emphatic reply.

CHAPTER 7

This Flowing World

"Our first naïve impression of Nature and matter is that of continuity. Be it a piece of metal or a volume of liquid, we invariably conceive it as divisible into infinity, and ever so small a part of it appears to us to possess the same properties as the whole."
—David Hilbert

In mathematics all roads lead back to Greece.

Here I am about to sketch the evolution of the idea of the infinitesimal. The place where the concept matured is Western Europe, and the time the seventeenth and the eighteenth centuries; yet when I endeavor to trace the origin of the idea I see another place and another time: the scene shifts back to classical Greece and to the memorable days of Plato.

The problem of the infinite, like the closely related problem of irrationals, grew up on Greek soil. There also occurred its first crisis, and it has had many since. The crisis came in the days of Plato, but it was not of Plato's making. Nor had the other orthodox philosophers of Greece any claim to having raised the issue. It was precipitated by a school of thinkers whom the leading philosophers of the period contemptuously called the "Sophists."

"Eleates" was the other name by which the orthodox thinkers stamped these obscure men, implying, perhaps, that their teachings were just as outlandish and insignificant as the homeland of their chief representative, Parmenides and Zeno. For Elea was a poor Greek colony in Southern Italy, "possessed of no other importance," says Laërius, "than the knowledge of how to raise virtuous citizens." To us, however, in retrospect, it seems that the Sophists were Elea's only claim to fame.

"The Arguments of Zeno of Elea have, in one form or another," says Russell, "afforded grounds for almost all the theories of space and time and infinity which have been constructed from his day to our own." Yet we don't know today whether these arguments were presented in the course of a debate or whether they appeared in the form of a book. Perhaps both! For we read in Plato's dialogue "Parmenides," one of the few sources we have on the obscure subject, of a visit which Zeno made to Athens in the company of his master, Parmenides. There is reference there to a previous visit during which, it appears, Zeno had presented his arguments. Yet when asked about these, Zeno replies:

> "Zeal for my master led me to write the book in the days of my youth, but one stole the writing; and, therefore, I had no choice whether it should become public; the motive for writing it was not the ambition of an older man, but the pugnacity of a young one."

Be this as it may, we know of the arguments only through Aristotle. Could the Stagyrite have resisted the temptation to distort the arguments of a dead adversary?

The rendition of the arguments in modern language is very difficult. Not that there is a dearth of translations—quite the contrary: we are suffering here from an *embarras du choix*. There are scores of translations and hundreds of paraphrases, and as for interpretations, no obscure passage in the Scriptures has been more honored. Each rendition reflects its author's pet theory, and there are almost as many theories as there are authors. The four Arguments of Zeno as recorded by Aristotle in his *Physica* are:

The First Argument: Dichotomy:

"The first is the one on the non-existence of motion, on the ground that what moves must always attain the middle point sooner than the end point."

The Second Argument: Achilles and the Tortoise:

"The second is the so-called Achilles. It consists in this, that the slower will never be overtaken in its course by the quicker, for the pursuer must always come first to the point from which the pursued has just departed, so the slower must necessarily be always still more or less in advance."

The Third Argument: The Arrow:

"If everything, when it is behaving in a uniform manner, is continually either moving or at rest, but what is moving is always in the now, then the moving arrow is motionless."

The Fourth Argument: The Stadium:

"The fourth is that concerning two rows, each row being composed of an equal number of bodies of equal size, passing each other on a race course, as they proceed with equal velocity in opposite directions; the one row originally occupying the space between the god and the middle point of the course, and the other that between the middle point and the starting point. This, he thinks, involves the conclusion that half a given time is equal to double the time."

Those who are metaphysically inclined see in the Arguments a refutation of the reality of motion. Others, like the historian Tannery, claim that Zeno had no such intention, but that, on the contrary, he used the undisputed reality of motion to point out the flagrant contradictions which reside in our notions of space, time, and continuity. Closely allied to this view is the opinion of Henri Bergson, who maintains that "the contradictions pointed out by the Eleatic school concern much less motion itself than the artificial reorganization of motion performed by our mind."

From this last point of view the value of the Arguments lies precisely in the fact that they forcefully bring out the position which mathematics occupies in the general scheme of human knowledge. The Arguments show that space and time and motion as perceived by our senses (or for this matter by their modern extensions, the scientific instruments) are not co-extensive with the mathematical concepts which bear the same name. The difficulties raised by Zeno are not of the type to alarm the pure mathematician—they do not disclose any logical contradictions, but only sheer ambiguities of language; the mathematician may dispose of these ambiguities by admitting that the symbolic world in which he creates is not identical with the world of his senses.

Thus the alleged properties of the straight line are of the geometer's own making. He deliberately disregards thickness and breadth, deliberately assumes that the thing common to two such lines, their point of intersection, is deprived of all dimension. Desirous of applying the laws of arithmetic to these geometrical beings, he admits, as we shall see, the validity of infinite processes, of which the infinte divisibility of a segment, the *dichotomy* of the Greeks, is but a particular instance. Classical geometry is a logical consequence of these assumptions, but the assumptions themselves are abitrary, a convenient fiction at best. The mathematician could reject the classical postulates, one or all, and substitute for them a new body of assumptions; he could, for instance, take for new elements the *stripe* and the *area* common to two stripes, and, calling these elements lines and points, construct a geometry altogether different from the classical doctrine, but just as consistent and perhaps just as fruitful.

But to the practical man, to the physicist, to the engineer, not all such systems are equally acceptable. The practical man demands an appearance of reality at least. Always dealing in the concrete, he regards mathematical terms not as symbols or thought but as images of reality. A system acceptable to the mathematician because of its inner consistency may appear to the practical man to be full of "contradictions" because of the incomplete manner in which it represents reality.

Strange though it may seem, it is the practical man who should be deeply concerned with the Arguments, because they attack the validity of the application of mathematics to physical reality. But, happily enough, the practical man is rarely interested in arguments.

The historical importance of the Arguments cannot be overestimated. For one thing, they forced the Greeks to adopt a new attitude towards the concept of time.

What Zeno substantially says in his first argument is this: The runner before reaching his goal must reach the midpoint of the course, and it takes him a *finite* time to achieve this. He also must reach the midpoint of the remaining distance, and this too will take a *finite* time. Now *what has been said once can always be repeated.* There are an infinite number of stages in his traversing of the race-course, and each one of these stages requires a finite time. But the sum of an infinite number of finite intervals is infinite. The runner will therefore never attain his goal.

Aristotle disposes of this argument as follows:

"Time and space are divided into the same and equal divisions. Wherefore also, Zeno's argument, that it is impossible to go through an infinite collection or to touch an infinite collection one by one in a finite time, is fallacious. For there are two senses in which the term 'infinte' is applied both to length and to time and in fact to all continuous things: either in regard to divisibility or in regard to number. Now it is not possible to touch things infinite as to number in a finite time, but it is possible to touch things infinite in regard to divisibility; for time itself is also infinite in this sense."

Thus the net result of the first two arguments (for the second is just an ingenious paraphrase of the first) is that it is impossible to assume *dichotomy of space* without simultaneously admitting *dichotomy of time.* But this is precisely what it is so difficult to grasp! For the divisibility of a line is easily conceived: we can readily materialize it by cutting a stick or marking a line. But "marking time" is just a figure of speech: time is the one thing on which we cannot experiment: it is either all in the past or all in the future. Dividing time into intervals was just an act of the mind to the Greeks, and is just an act of the mind to us.

Endowing time with the attribute of infinite divisibility is equivalent to representing time as a geometrical line, to identifying *duration* with *extension.* It is the first step towards the *geometrization* of mechanics. Thus the first arguments of Zeno were directed against the principle on which the four-dimensional world of modern Relativity is built.

The real punch of the Arguments was reserved for the last two; as though Zeno foresaw the defense of his opponents and prepared to meet it. The fourth, which contains the germ of the problem of Relativity, does not concern us here. It is the third argument that forcefully exposes the chasm between motion as perceived by our senses and the mathematical fiction which masquerades under the same name.

We can hear Zeno's answer in rebuttal:

> "You say that just as space consists of an infinity of contiguous points, so time is but an infinite collection of contiguous instants? Good! Consider, then, an arrow in its flight. At any instant its extremity occupies a definite point in its path. Now, while *occupying this position* it must be *at rest* there. But how can a point be motionless and yet in motion at the same time?"

The mathematician disposes of this argument by *fiat*: Motion? Why, motion is just a correspondence between position and time. Such a correspondence between variables he calls a *function*. The law of motion is just a function, in fact the prototype of all *continuous functions*. Not different in substance from the case of a cylinder filled with gas and provided with a piston which is free to slide within the cylinder. To every possible position of the piston there will correspond a definite pressure within the cylinder. To obtain the pressure corresponding to any position we may stop the piston in this position and read the pressure gauge.

But is it the same with a moving body? Can we stop it at any instant without curtailing the very motion which we are observing? Assuredly not! What is it then that we mean by the moving body's *occupying a certain position at a certain time?* We mean that while we cannot conceive of a physical procedure which will arrest the arrow in its flight without destroying the flight, there is nothing to prevent our doing so by an *act of the mind*. But the only reality behind this act of mind is that *another* arrow can be imagined as motionless at this point and at this instant.

Mathematical motion is just an infinite succession of states of rest, i.e., mathematics reduces dynamics to a branch of statics.

The principle that accomplishes this transition was first formulated by d'Alembert in the eighteenth century. This identification of motion with a succession of contiguous states of rest, during which the moving body is in equilibrium, seems absurd on the face of it. And yet motion made up of *motionless* states is no more, nor less absurd than length made up of *extensionless* points, or time made up of *durationless* instants.

True, this abstraction is not even the skeleton of the real motion as perceived by our senses! When we see a ball in flight we perceive the motion as a whole and not as a succession of infinitesimal jumps. But neither is a mathematical line the true, or even the fair, representation of a wire. Man has for so long been trained in using these fictions that he has come to prefer the substitute to the genuine article.

The subsequent course of Greek science shows clearly how great was the influence which the crisis precipitated by the Arguments of Zeno exercised on the mathematical thought of the Hellenes.

On the one hand this crisis ushered in an era of sophistication. It was the natural reaction from the naïve verbiage of the Pythagoreans, that strange mixture of mathematical ideas with religious slogans and vague metaphysical speculations. What a contrast to this is the sever rigor of Euclid's *Elements*, which to this day serves as a model for mathematical disciplines!

On the other hand, by instilling into the mind of the Greek geometers the *horror infiniti*, the Arguments had the effect of a partial paralysis of their creative imagination. The infinite was taboo, it had to be kept out, at any cost; or, failing in this, camouflaged by arguments *ad absurdum* and the like. Under such circumstances not only was a positive theory of the infinite impossible, but even the development of infinite processes, which had reached quite an advanced stage in pre-Platonic times, was almost completely arrested.

We find in classical Greece a confluence of most fortunate circumstances: a line of geniuses of the first rank, Eudoxus, Aristarchus, Euclid, Archimedes, Apollonius, Diophantus, Pappus; a body of traditions which encouraged creative effort and speculative thought and at the same time furthered a critical spirit, safeguarding the investigator against the pitfalls of an ambitious imagination; and finally, a social structure most propitious to the development of a leisure class, providing a constant flow of thinkers, who could devote themselves to the pursuit of ideas without regard to immediate utility—a combination of circumstances, indeed, which is not excelled even in our own days. Yet Greek mathematics stopped short of an algebra in spite of a Diophantus, stopped short of an analytic geometry in spite of an Apollonius, stopped short of an infinitesimal analysis in spite of an Archimedes. I have already pointed out how the absence of a notational symbolism thwarted the growth of Greek mathematics; the *horror infiniti* was just as great a deterrent.

In the *method of exhaustion,* Archimedes possessed all the elements essential to an infinitesimal analysis. For modern analysis is but the theory of infinite processes, and infinite processes have for foundation the idea of *limit.* The precise formulation of this idea I reserve for the next chapter. It is sufficient to say here that the idea of limit as conceived by Archimedes was adequate for the development of the calculus of Newton and Leibnitz and that it remained practically unchanged until the days of Weierstrass and Cantor. Indeed the *calculus of limits* rests on the notion that two variable magnitudes will approach a state of equality if their difference could be made deliberately small, and this very idea is also the basis of the method of exhaustion.

Furthermore, the principle provides an actual method for determining the limit. This consists in "trapping" the variable magnitude between two others, as between the two jaws of a vise. Thus in the case of the periphery of the circle, of which I have already spoken, Archimedes grips the circumference between two sets of regular polygons of an increasing number of sides, of which one set is circumscribed to the circle and the other is inscribed in it. As I said before, Archimedes showed by this method that the number π is contained between $3\frac{1}{7}$ and $3\frac{10}{71}$. By this method he also found that the area under a parabolic arch is equivalent to two-thirds of the area of a rectangle of the same base and altitude—the problem which was the precursor of our modern integral calculus.

Yes, in all justice it must be said that Archimedes was the founder of infinitesimal analysis. What the method of exhaustion lacked of being the integral calculus of the eighteenth century was a proper symbolism, and a positive—or, shall I say, a naïve—attitude towards the infinite. Yet no Greek followed in the footsteps of Archimedes, and it was left to another epoch to explore the rich territory discovered by the great master.

When, after a thousand-year stupor, European thought shook off the effect of the sleeping powders so skillfully administered by the Christian Fathers, the problem of infinity was one of the first to be revived.

Characteristic of this revival, however, was the complete absence of the critical rigor of the Greeks, and this in spite of the fact that Renaissance mathematics relied almost entirely on Greek sources. The rough-and-ready methods inaugurated by

Kepler and Cavalieri were continued, with only a pretense of refinement, by Newton and Leibnitz, by Wallis, the inventor of the symbol for infinity, by the four Bernoullis, by Euler, by d'Alembert.

They dealt with infinitesimals as fixed or variable according to the exigencies of the argument; they manipulated infinite sequences without much rhyme or reason; they juggled with limits; they treated divergent series as if these obeyed all rules of convergence. They defined their terms vaguely and used their methods loosely, and the logic of their arguments was made to fit the dictates of their intuition. In short, they broke all the laws of rigor and of mathematical decorum.

The veritable orgy which followed the introduction of the infinitesimals, or the *indivisibilia,* as they were called in those days, was but a natural reaction. Intuition had too long been held imprisoned by the severe rigor of the Greeks. Now it broke loose, and there were no Euclids to keep its romantic flight in check.

Yet another cause may be discerned. It should be remembered that the brilliant minds of that period were all raised on scholastic doctrine. "Let us have a child up to the age of eight," said a Jesuit once, "and his future will take care of itself." Kepler reluctantly engaged in astronomy after his hopes of becoming an ecclesiastic were frustrated; Pascal gave up mathematics to become a religious recluse; Descartes's sympathy for Galileo was tempered by his faith in the authority of the Church; Newton in the intervals between his masterpieces wrote tracts on theology; Leibnitz was dreaming of number schemes which would make the world safe for Christianity. To minds whose logic was fed on such speculations as Sacrament and Atonement, Trinity and Trans-substantiation, the validity of infinite processes was a small matter indeed.

This may be taken as a rather belated retort to Bishop Berkeley. A quarter of a century after the publication of Newton's epoch-making work on the infinitesimal calculus, the bishop wrote a tract entitled: "The Analyst; a Discourse Addressed to an Infidel Mathematician." The contention that too much is taken on faith in matters of religion, the bishop counters by pointing out that the premises of mathematics rest on no securer foundation. With inimitable skill and wit he subjects the doctrine of infinitesimals to a searching analysis and discloses a number of loose arguments, vague statements, and glaring contradictions. Among these are the terms "fluxion" and "difference"; and against these the bishop directs the shafts of his splendid Irish humor: "He who can digest a second or third fluxion, a second or third difference, need not, methinks, be squeamish about any point in Divinity."

The "fluxions" of Newton, the "differences" of Leibnitz, are today called *derivatives* and *differentials*. They are the principal concepts of a mathematical discipline which, together with analytical geometry, has grown to be a powerful factor in the development of the applied sciences: the *Differential and Integral Calculus*. Descartes is credited with the creation of analytic geometry; the controversy as to whether it was Newton or Leibnitz who first conceived the calculus raged throughout the eighteenth century and is not quite settled even today. And yet, we find the principles of both disciplines clearly indicated in a letter which Fermat addressed to Roberval, dated October 22, 1636, a year before Descartes's *Geometria* appeared, and sixty-eight years before the publication of Newton's *Principia*. If it were not for Fermat's unaccountable habit of not publishing his researches, the creation of both analytic geometry and the calculus would have been credited to this Archimedes of the Renaissance, and the mathematical world would have been spared the humiliation of a century of nasty controversy.

The substance of Newton's principle can be illustrated by the example of motion, which, incidentally, was the first subject to which the differential calculus was applied. Consider a particle in motion along a straight line. If in equal times equal spaces are covered, then the particle is said to move *uniformly;* and the distance covered in a unit of time, say a second, is called the *velocity* of this uniform motion. Now if the distances covered in equal intervals of time are not equal, i.e., if the motion is *non-uniform,* there is no such thing as velocity in the sense in which we have just used the word. Yet we may divide the distance which was covered in a certain interval by the time interval and call this ratio the *average* velocity of the particle in this interval. Now it is this ratio that Newton would call *prime* ratio. This number, however, obviously depends on the length of the interval considered. However, notice that the smaller the interval the closer does the velocity approach a certain fixed value. ... We have here an example of *a sequence* in which the difference between succeeding terms is growing continually less until after a while two contiguous terms will become indistinguishable. Now let us conceive (and such a conception is justified by our intuitive notion of the *continuity of space and time)* that we continue diminishing the interval of time *indefinitely.* Then, the ultra-ultimate term of the sequence (the *ultima ratio* of Newton) will, according to Newton, represent the *velocity at the point at the beginning of the interval.*

Today we say: *by definition* the velociy of the moving point at any time is the limiting value of the average velocity when the interval to which the average velocity pertains diminishes indefinitely. In the days of Newton they were not so careful.

The ultimate ratios were also called by Newton *fluxions*. The fluxion was the rate of change of a variable magnitude, such as length, area, volume, pressure, etc. These latter Newton called the *fluents*. It is to be regretted that these expressive words were not retained, but were replaced by such indifferent terms as *derivative* and *function*. For the Latin *fluere* means "to flow"; *fluent* is "the flowing," and *fluxion* "the rate of flow."

Newton's theory dealt with continuous magnitudes and yet postulated the infinite divisibility of space and time; it spoke of a flow and yet dealt with this flow as if it were a succession of minute jumps. Because of this, the theory of fluxions was open to all the objections that two thousand years before had been raised by Zeno. And so the age-long feud between the "realists," who wanted a mathematics to comply with the crude reality of man's senses, and the "idealists," who insisted that reality must conform to the dictates of the human mind, was ready to be resumed. It only awaited a Zeno, and the Zeno appeared in the strange form of an Anglican ecclesiastic. But let me leave the word to George Berkeley, later Bishop of Cloyne:

"Now, as our Sense is strained and puzzled with the perception of objects extremely minute, even so the Imagination, which faculty derives from Sense, is very much strained and puzzled to frame clear ideas of the least particles of time, or the least increments generated therein; and much more so to comprehend the moments, or those increments of the flowing quantities *in statu nascenti*, in their very first origin or beginning to exist, before they become finite particles. And it still seems more difficult to conceive the abstract velocities of such nascent imperfect entities. But the velocities of the velocities—the second, third, fourth, and fifth velocities, etc.—exceed, if I mistake not, all human understanding. The further the mind analyseth and pursueth these fugitive ideas the more it is lost and bewildered; the objects, at first fleeting and minute, soon vanishing out of sight. Certainly, in any sense, a second or a third fluxion seems an obscure Mystery. The incipient celerity of an incipient celerity, the nascent augment of a nascent augment, i.e., of a thing which hath no magnitude—take it in what light you please, the clear conception of it will, if I mistake not, be found impossible ...

"The great author of the method of fluxions felt this difficulty, and therefore he gave in to those nice abstractions and geometrical metaphysics without which he saw nothing could be done on the received principles. ...It must, indeed, be acknowledged that he used fluxions like the scaffold of a building, as things to be laid aside or got rid of as soon as finite lines were found proportional to them. But then these finite exponents are found by the help of fluxions....And what are these fluxions? The velocities of evanescent increments. And what are these same evanescent increments? They are neither finite quantities, nor quantities infinitely small, nor yet nothing. May we not call them the ghosts of departed quantities? ...

"And, to the end that you may more clearly comprehend the force and design of the foregoing remarks, and pursue them still farther in your own meditations, I shall subjoin the following Queries. ...

"*Query 64.* Whether mathematicians, who are so delicate in religious points, are strictly scrupulous in their own science? Whether they do not submit to authority, take things upon trust, and believe points inconceivable? Whether they have not their mysteries, and what is more, their repugnances and contradictions?"

And the net result of Berkeley's witty perorations? Well, in so far as it attacked inaptness and inconsistency in the mathematical terminology, it performed a genuine service. Succeeding decades saw a considerable change: such words as prime and ultimate, nascent and incipient, fluent and fluxion, were abandoned. The *indivisibilia* became the *infinitesimals* of today; the infinitesimal being merely a variable quantity that approaches zero as a limit. The whole situation became slowly but surely dominated by the central idea of limit.

Had Bishop Berkeley reappeared fifty years after he wrote "The Analyst" he would not have recognized the child he had scolded, so modest had it become. But would he have been satisfied? Not Berkeley! For the sharp eyes of the acute bishop would have detected the same leopard behind the changed spots. What he had objected to was not so much the lack of conciseness in language (although this too came in for its share in his critique); but rather what Zeno had pointed out: the failure of the new method to satisfy our intuitive idea of the continuous as of something uninterrupted, something indivisible, something that had no parts, because any attempt to sever it into parts would result in the destruction of the very property under analysis.

And if we strain our imaginations still more and imagine the bishop re-appearing in our own midst, we would hear him raising the same objections, leveling the same accusations. But this time to his surprise and delight he would find in the enemy camp a powerful party of men who would not only defend him but hail him as a pioneer.

But of this later.

And in the meantime analysis grew and grew, not heeding the warnings of the critics, constantly forging ahead and conquering new domains. First geometry and mechanics, then optics and acoustics, propagation of heat and thermodynamics, electricity and magnetism, and finally even the laws of the Chaos came under its direct sway.

Says Laplace:

"We may conceive the present state of the universe as the effect of its past and the cause of its future. An Intellect who at any given instant knew all the forces that animate nature and the mutual position of the beings who compose it, were this Intellect but vast enough to submit his data to analysis, could condense into a single formula the movement of the greatest body in the universe and that of the lightest atom; to such an Intellect nothing would be uncertain, for the future, even as the past, would be ever present before his eyes."

And yet this magnificent structure was created by the mathematicians of the last few centuries without much thought as to the foundations on which it rested. Is it not remarkable then, that in spite of all the loose reasoning, all the vague notions and unwarranted generalization, so few serious errors had been committed? "Go ahead, faith will follow" were the encouraging words with which d'Alembert kept reinforcing the courage of the doubters. As though heeding his words, they did forge ahead, guided in their wanderings by a sort of implicit faith in the validity of infinite processes.

Then came the critical period: Abel and Jacobi, Gauss, Cauchy and Weierstrass, and finally Dedekind and Cantor, subjected the whole structure to a searching analysis, eliminating the vague and ambiguous. And what was the net result of this reconstruction? Well, *it condemned the logic of the pioneers, but vindicated their faith.*

The importance of infinite processes for the practical exigencies of technical life can hardly be overemphasized. Practically all applications of arithmetic to geometry, mechanics, physics and even statistics involve these processes directly or indirectly.

Indirectly because of the generous use these sciences make of irrationals and transcendentals; directly because the most fundamental concepts used in these sciences could not be defined with any conciseness without these processes. Banish the infinite process, and mathematics pure and applied is reduced to the state in which it was known to the pre-Pythagoreans.

Our notion of the length of an arc of a curve may serve as an illustration. The physical concept rests on that of a bent wire. We imagine that we have *straightened* the wire without *stretching* it; then the segment of the straight line will serve as the measure of the length of the arc. Now what do we mean by "without stretching"? We mean without a change in length. But this term implies that we already know something about the length of the arc. Such a formulation is obviously a *petitio principii* and could not serve as a mathematical definition.

The alternative is to inscribe in the arc a sequence of rectilinear contours of an increasing number of sides. The sequence of these contours approaches a limit, and the length of the arc is defined as the limit of this sequence.

And what is true of the notion of length is true of areas, volumes, masses, moments, pressures, forces, stresses and strains, velocities, accelerations, etc., etc. All these notions were born in a *"linear," "rational"* world where nothing takes place but what is straight, flat, and uniform. Either, then, we must abandon these elementary rational notions—and this would mean a veritable revolution, so deeply are these concepts rooted in our minds; or we must adapt those rational notions to a world which is neither flat, nor straight, nor uniform.

But how can the flat and the straight and the uniform be adapted to its very opposite, the skew and the curved and the non-uniform? Not by a finite number of steps, certainly! The miracle can be accomplished only by that miracle-maker the *infinite*. Having determined to cling to the elementary rational notions, we have no other alternative than to regard the "curved" reality of our senses as the ultra-ultimate step in an infinite sequence of *flat* worlds which exist only in our imagination.

The miracle is that it works!

CHAPTER 8

The Art of Becoming

"No more fiction for us: we calculate; but that we may calculate, we had to make fiction first."

—Nietzsche

Returning now to irrationals, I shall endeavor to show the close connection that exists between this problem and the problem of continuity which I discussed in the preceding chapter. But first let me resume the situation where I left it before taking up the problem of continuity.

The attempt to apply rational arithmetic to a problem in geometry resulted in the first crisis in the history of mathematics. The two relatively simple problems, the determination of the diagonal of a square and that of the circumference of a circle, revealed the existence of new mathematical beings for which no place could be found within the rational domain. The inadequacy of rational arithmetic was thus forcefully brought home.

A further analysis showed that the procedures of algebra were generally just as inadequate. So it became apparent that an extension of the number field was unavoidable. But how was this to be accomplished? How can an infinity, nay, an infinite variety of infinite collection of irrationals, be inserted into the closely knit texture of the rational aggregate of numbers?

We must recast the old number concept—so much is certain. And since the old concept failed on the terrain of geometry, we must seek in geometry a model for the new. The continuous indefinite straight line seems ideally adapted for such a model. Here, however, we strike a new difficulty: if our number domain is to be identified with the line, then to any individual number must correspond a point. But what is a point? We must have, if not a definition, at least a clear-cut idea of what we mean by an element of a line, *a point.*

Now, the general notion of a point as a geometrical being without dimension is, of course, a fiction; but when we analyze this fiction we find that back of it are three distinct ideas. In the first place, we conceive the point as a sort of *generating* element which in its motion describes the line. This idea seems to fit best our intuitive idea of *continuity,* which is the first attribute we ascribe to the line. When, however, we attempt to take this dynamic conception as a basis for the analogy between the line and the number domain, we find the two incompatible.

Indeed, our senses perceive motion as something individual, indivisible, uninterrupted. The very act of resolving motion into elements results in the destruction of the continuity which we have resolved to preserve. For the sake of number, it is necessary to regard the line as a succession of infinitesimal resting-stations, and this is repugnant to the very idea of motion conceived by us as the direct opposite of rest. Therein lies the force of Zeno's arguments.

We saw how the mathematician tried to bridge the discrepancy by the invention of the infinitesimal analysis; we saw how this analysis, starting with geometry and mechanics, succeeded in acquiring a dominant position in every field of the exact sciences until it became a veritable mathematical *theory of change.*

Speaking pragmatically, this sweeping triumph of analysis is sufficient proof of the validity of its methods. But while it may be true that the proof of the pudding is in the eating, the eating sheds no light on what a pudding is. The very success of analysis only accentuates the age-old question: what constitutes a continuum?

In the second place, the point may be regarded as an intersection of two lines, i.e., as a *mark left* on the line in question by another line. As such, it is just a *partition*, a manner of severing the line into two mutually exclusive complementary regions. It is this idea which Richard Dedekind took as a point of departure in his epoch-making essay entitled "Continuity and Irrational Numbers," which appeared in 1872. Of this I shall tell in the next chapter.

Finally, we may regard the point as a *limiting position* in an *infinite process* applied to a segment of a line. This process may take many forms; for the present it is sufficient to say that a typical instance is the dichotomy of the Greeks. The arithmetical counterpart of the infinite process is the *infinite sequence,* and it is the rational infinite sequence that Georg Cantor uses as the vehicle for his famous theory of irrationals, which was first published in 1884. It is this simple and far-reaching idea that is the subject of this chapter.

A sequence is *rational* if its terms are all rational numbers; it is *infinite* if every term in it has a successor. A set of operations generating an infinite sequence I shall call an infinite process.

The prototype of all infinite processes is *repetition*. Indeed, our very concept of the infinite derives from the notion that *what has been said or done once can always be repeated.* When repetition is applied to a rational number *a*, we obtain the *repeating* sequence

$$a, a, a, a, \ldots$$

I shall say that this sequence *represents the number a.*

Another fundamental operation, which I shall call the *serial process*, is that of successive addition. Given the sequence

$$a, b, c, d, e, f, g \ldots,$$

the serial process creates a new sequence

$$a, a + b, a + b + c, a + b + c + d, \ldots,$$

which we call the *series* generated by the sequence *a, b, c*, ... Thus from the repeating sequence 1, 1, 1, ... we derive the *natural sequence 1, 2, 3, 4* ...

The serial process can obviously be applied to any sequences; and therefore to every sequence corresponds a series. Of the greatest importance, however, are those series which are generated by *evanescent sequences*. These latter are characterized by the gradual diminution of the successive terms, so that it is possible to "go out" sufficiently far to find terms less in value than any assignable number. Of this type are the sequences

$$\tfrac{1}{2}, \tfrac{1}{4}, \tfrac{1}{8}, \tfrac{1}{16}, \tfrac{1}{32} \ldots$$

$$\tfrac{1}{2}, \tfrac{1}{3}, \tfrac{1}{4}, \tfrac{1}{5}, \tfrac{1}{6} \ldots$$

Now, given any two sequences, a third sequence can be formed by subtracting one from the other, term by term. It may happen that the *difference sequence* so derived is evanescent, as is the case with the two sequences

$$\tfrac{2}{1}, \tfrac{3}{2}, \tfrac{4}{3}, \tfrac{5}{4}, \tfrac{6}{5}, \tfrac{7}{6} \ldots, \text{ and}$$

$$\tfrac{1}{2}, \tfrac{2}{3}, \tfrac{3}{4}, \tfrac{4}{5}, \tfrac{5}{6}, \tfrac{6}{7} \ldots$$

Here the differences between corresponding terms form the sequence

$$\frac{3}{1\times2},\frac{5}{2\times3},\frac{7}{3\times4},\frac{9}{4\times5},\frac{11}{5\times6},\frac{13}{6\times7},\frac{15}{7\times8},\ldots$$

The denominator of each term is the product of two successive numbers, and the numerator is the sum of the same numbers. The one-thousandth term of this sequence is less than .002, the one-millionth is less than .000002, etc. The sequence is certainly evanescent.

I shall call two sequences whose difference is evanescent, *asymptotic.* Now, one of the two asymptotic sequences may be a repeating sequence, as for instance in the case of

$$1, 1, 1, 1, \ldots$$
$$\frac{1}{2}, \frac{2}{3}, \frac{3}{4}, \frac{4}{5}, \ldots$$

The repeating sequence represents the rational number 1. I shall say that the second sequence asymptotic to the first *also represents the number 1,* or that it *converges towards 1 as a limit.*

It stands to reason that if two sequences are asymptotic to a third, they are asymptotic to each other, and, furthermore, if one converges to a certain rational number as a limit, the same is true of the other. Now since this is so, a great number of sequences may, in spite of their difference in form, represent the same number. And such is, indeed, the case. Thus the number 2 is capable of an infinite number of representations by rational sequences, of which a few are:

$$1.9, 1.99, 1.999, 1.9999 \ldots$$
$$2.1, 2.01, 2,001, 2,0001 \ldots$$
$$1\frac{1}{3}, 1\frac{2}{3}, 1\frac{3}{4}, 1\frac{4}{5} \ldots$$
$$1\frac{1}{2}, 1\frac{3}{4}, 1\frac{7}{8}, 1\frac{15}{16} \ldots$$

The same is true of any rational number. In particular, *any evanescent sequence may be regarded as a representation of the rational number 0.*

The simplest type of sequence, and one that is at the same time of great historical and theoretical importance, is the *geometrical sequence*. Here, having selected any number for the first term, and any other number for ratio, we proceed from term to term by multiplication through the ratio. Any repeating sequence may be regarded as a special geometrical sequence, the multiplying ratio here being 1. If we eliminate this trivial case, we can classify geometrical sequence into *increasing* and *diminishing*. Examples of these are

$$2, 4, 8, 16, 32, 64, \ldots 2^n, \ldots \text{and}$$
$$1, \tfrac{1}{3}, \tfrac{1}{9}, \tfrac{1}{27}, \tfrac{1}{81} \ldots \ldots, \tfrac{1}{3}^n \ldots$$

In the increasing geometrical sequence, the terms grow *indefinitely* in absolute value; that is to say, if we "go out" far enough, we can find terms which exceed any assignable number, no matter how great. Such sequences are said to *diverge*.

The diminishing sequence is always evanescent and for this reason is of special interest to us here. But what makes it particularly valuable is the fact that the series generated by such an evanescent geometrical sequence will always converge towards a rational limit, and that conversely *any rational number can be regarded as the limit of some rational geometrical series.* Moreover, here we have one of the rare cases where the "sum of a series" can actually be evaluated in terms of the immediate data.

The series generated by a geometrical sequence is called a *geometrical progression*. An evanescent geometrical sequence generates a *convergent progression*. If the sequence begins with the term *a* and has *r* for ratio, the limit is given by the simple formula

$$S = \frac{a}{1-r}$$

This limit is called the *sum* of the progression.

The dichotomy sequence in the first argument of Zeno is the geometrical progression

$$\tfrac{1}{2}, \tfrac{1}{4}, \tfrac{1}{8}, \tfrac{1}{16}, \ldots$$

It generates the series

$$\tfrac{1}{2}, \tfrac{3}{4}, \tfrac{7}{8}, \tfrac{15}{16}, \ldots.$$

This latter converges towards 1, as can be seen directly or by the summation formula.

The sum

$$\tfrac{1}{2} + \tfrac{1}{4} + \tfrac{1}{8} + \tfrac{1}{16} + \ldots.$$

represents the finite number 1 in spite of the argument of Zeno that this sum is extended over an infinite number of terms. The introduction of the concepts of convergence and limits may be objected to on one ground or another, but once accepted, the argument of Zeno, that a sum of an infinite series of numbers must by necessity be infinite, loses its force.

The second Zeno argument also involves a geometrical progression. To deal with it concretely, let us assume that Achilles advances at the rate of 100 feet per minute, and the tortoise at the rate of 1. When will Achilles overtake the tortoise, if the original handicap was 990 feet? Never, says Zeno. "Common sense" tells us, however, that Achilles gains on the tortoise 99 feet per minute, and that the original distance of 990 feet will be wiped out at the end of the first 10 minutes. But let us argue in the Zeno fashion. By the time Achilles reaches the position originally held by the tortoise, the tortoise has advanced 1/100 of the handicap, or 9.9

feet; by the time Achilles arrives at this second position, the tortoise has advanced 1/100 of 9.9 or 0.099 feet. But "what was said once can always be repeated." The handicap is being reduced in the geometrical progression

$$990, 9.9, 0.099, 0.00099, ...,$$

the sum of which by the summation formula is 1000. Achilles will have covered 1000 feet before overtaking the tortoise, and this will take him 10 minutes. Again, the sum of an infinite number of terms *may be finite.*

Periodic decimal fractions are but geometrical series in disguise.

Consider for instance the infinite decimal fraction, which is of the *pure* periodic type,

$$0.36363636 ...$$

I shall write this short 0.(36)
The actual meaning of this is

$$36/100 + 36/10,000 + 36/1,000,000, ...$$

This, however, is a geometrical series of ratio 1/100, and the summation formula shows that the series converges towards the rational limit 36/99 or 4/11. The same is true of a so-called *mixed* periodic fraction, such as 0.34 (53), for instance: for multiplying this by 100 we obtain the pure periodic fraction $34\frac{53}{99}$, and since the mixed periodic fraction is 1/100 of this, we have 0.34(53) = 3419/9900.

Even a terminating fraction may be regarded as a periodic fraction of *period zero.* For instance:

$$2.5 = 2.50000 ... = 2.5(0)$$

Now we were taught in school how to convert any common fraction into a decimal. The procedure is called long division, and we know by experience that it leads either to a terminating fraction, as in the case of 1/8 which is equivalent to 0.125, or to an infinite periodic, as in the case of 1/7, which is represented by 0.(142857). This property can be proved in all rigor and may be formulated in the statement: Any *rational number can be represented in a unique way as an infinite periodic decimal fraction; conversely, any periodic decimal fraction represents a rational number.*

On the other hand, we can obviously construct any number of decimal sequences which *while infinite* are *not periodic.* The distribution of the digits may be chaotic, or it may follow a regular yet *non-periodic* law. Such, for instance, is the case of the decimal series

$$1.10111213 \ldots 192021 \ldots 100101 \ldots.$$

If we could find a repeating rational sequence $a, a, a, \ldots$, which would be asymptotic to this decimal sequence, then this latter would represent the rational number a. But we know that this is impossible; for if it were possible, the sequence would be periodic and such is not the case. What then does this series represent? We do not know. The manner in which we defined convergence and limit precludes all possibility of classifying this sequence as a number. And yet, there are our intuitive notions of convergence and of limit as of something *growing* and yet never exceeding a certain magnitude, or as of something *waning* and yet never falling below a certain magnitude. From this intuitive viewpoint the infinite non-periodic decimal series *does converge,* and the same is true of many other sequences, such as

$$(1\tfrac{1}{2})^2, (1\tfrac{1}{3})^3, (1\tfrac{1}{4})^4, (1\tfrac{1}{5})^5 \ldots,$$

which, incidentally, represents the transcendental e.

It is this naïve idea of convergence and limit that was accepted as axiomatic in the early days of analysis, and it must be admitted that, in spite of the pitfalls to which it led, it is to this idea that the calculus owed its first successes. Thus the questions which present themselves most naturally to the mind are these: Is it possible to clothe this vague intuitive idea of convergence and limit in a precisely formulated definition? Is it possible by such a definition to create a new instrument which will permit us to deal with these new mathematical beings, exhibited by the non-periodic decimal series and other sequences, with the same certainty as when dealing with the special kind which admit rational limits?

To answer these questions, we must examine whether among the properties of the special sequences which converge towards rational limits there exists one, which would permit of an immediate generalization to the vastly more extended kind which do not so converge. Georg Cantor discovered such a property in what I shall call the *self-asymptotic nature* of a convergent sequence.

To exhibit this, let us consider again the dichotomy series. Let us "advance" this sequence by curtailing the first term and making the second term first, the third second, etc. This advancing process will create the succession of sequences

$$\tfrac{1}{2}, \tfrac{3}{4}, \tfrac{7}{8}, \tfrac{15}{16}, \tfrac{31}{32}, \tfrac{63}{64}, \tfrac{127}{128}, \ldots\ldots\ldots\ldots$$

$$\tfrac{3}{4}, \tfrac{7}{8}, \tfrac{15}{16}, \tfrac{31}{32}, \tfrac{63}{64}, \tfrac{127}{128}, \tfrac{255}{256}, \ldots\ldots\ldots\ldots$$

$$\tfrac{7}{8}, \tfrac{15}{16}, \tfrac{31}{32}, \tfrac{63}{64}, \tfrac{127}{128}, \tfrac{255}{256}, \tfrac{511}{512}, \ldots\ldots\ldots$$

which obviously can be continued indefinitely. Now even a casual examination of these sequences shows that they are all asymptotic to each other; that is, that the difference sequence formed with any two is evanescent.

It can be shown that this *self-asymptotic* property holds for all those sequences which converge towards a rational limit; but it is by no means confined to these: in fact, any infinite non-periodic decimal series has also the same property. Indeed, consider as an example the decimal series

.101112131415 ...,

which can be written as

.1, .10, .101, .1011, .10111, .101112, ...

It is obvious that curtailing any number of these rational approximations will not affect even the appearance of the sequence. Thus, we can write it in the form:

.101112, .1011121, .10111213, .101112131, ...

which is certainly asymptotic to the first.

And so Cantor extended the idea of convergence, which hitherto applied only to those sequences which were asymptotic to some rational repeating sequence, by identifying the two terms *self-asymptotic* and *convergent*. Moreover, he extended the idea of *limit* by regarding the self-asymptotic sequence as generating a new type of mathematical being which he identified with what had long before him been called *real number*.

Now, applying the name *number* to such beings would be justifiable if it could be shown that all the stipulations of the principle of permanence were satisfied.

That the first stipulation is satisfied follows from the fact that among the convergent sequences there are those which admit rational numbers for limits. The second stipulation implies the criteria of rank. Let us then consider two sequences (A) and (B) which define the two real numbers *a* and *b*, and let us form the difference sequence (A – B). It may happen that this latter is evanescent, in which case (A) and (B) are asymptotic, and we say then that the numbers a and b are *equal*. As an example of this take the two sequences

$$(1\tfrac{1}{2})^2 \qquad (1\tfrac{1}{3})^3 \qquad (1\tfrac{1}{4})^4 \qquad (1\tfrac{1}{5})^5 \dots,$$

and

$$2 + \tfrac{1}{2}!, \, 2 + \tfrac{1}{2}!, \, + \tfrac{1}{3}!,$$
$$2 + \tfrac{1}{2}! + \tfrac{1}{3}! + \tfrac{1}{4}!,$$
$$2 + \tfrac{1}{2}! + \tfrac{1}{3}! + \tfrac{1}{4}! + \tfrac{1}{5}!, \dots$$

These can be proved to be asymptotic and therefore represent the same real number, which is the transcendental e.

If the difference sequence is not evanescent, then it may happen that, beginning with a certain term, all terms are positive, in which case we say that the sequence (A) *dominates* the sequence (B), or that the real number a is *greater* than the real number b. And again, if beginning with a certain term all the succeeding terms of the difference sequence are negative, then (A) is *dominated* by (B), and we then say that a is *less* than b. These criteria, it can be shown, reduce to the standard when (A) and (B) admit rational limits.

Finally, we define the sum and the product of two real numbers as the real numbers defined by sequences obtained by adding or multiplying the corresponding terms of these sequences. This, of course, implies that these resultant sequences are themselves convergent, and that such is indeed the case can be proved in all rigor. Furthermore, it can be shown that addition and multiplication so defined are associative, commutative, and distributive.

From the standpoint of the principle of permanence these new magnitudes can therefore be admitted as full-fledged numbers. By their adjunction the rational field becomes but a province in a vastly more extensive realm, which we shall call the *domain of real numbers.*

Will this new domain contain the irrationals of algebra, the transcendental of analysis? Yes, and to exhibit this let us return to the equation $x^2 = 2$, which over two thousand years ago, in the guise of the problem to determine the diagonal of a square, precipitated the crisis which has now culminated in the erection of the real number domain.

We were taught in school an algorithm for the extraction of the square root. This procedure gives for what we call $\sqrt{2}$ a set of rational approximations which form the convergent sequence

$$1, 1.4, 1.41, 1.414, 1.4142, 1.41421, \ldots.$$

This sequence has no rational limit, but the sequence obtained by squaring each term in it, namely,

$$1, 1.96, 1.9881, 1.999396, \ldots,$$

converges towards the rational number 2.

Therefore when we say that the positive solution of the equation $x^2 = 2$ is the sequence in question and denote the number defined by it $\sqrt{2}$, we mean not only that the sequence of squares converges, but that it belongs to that rare type of convergent sequences which possesses a *rational limit,* in our case the number 2. In other words, the sequence represents $\sqrt{2}$ because we admit that the number 2, while not a perfect square, is the limit towards which a sequence of perfect squares converges.

A similar procedure will apply to other algebraic or transcendental equations. The actual discovery of an algorithm, which would in any particular instance generate the sequence that converges towards the sought solution, may be a matter of considerable mathematical difficulty. Yet, once devised, the sequence can always be paraphrased into an infinite decimal fraction which by its very nature is convergent and, therefore, represents a real number.

Thus, the admission of the validity of infinite processes takes us out of the restricted boundaries of rational arithmetic. It creates a *general arithmetic, the arithmetic of real numbers,* and this furnishes us with the means of attacking problems before which rational arithmetic stood powerless.

It would seem at first that by giving the very general name *real* to the limits of rational sequences we lacked foresight. For, indeed, it is only natural now to consider infinite sequences of such irrationals. If we had called the first type irrationals of the first *rank,* these new limits could then be called irrationals of the second rank; from these we would derive irrationals of the third rank, etc., etc. That this is no idle juggling is testified by such a simple expression as $\sqrt{1+\sqrt{2}}$, the direct interpretation of which creates the *irrational* sequence.

$$\sqrt{2.4}, \ \sqrt{2.41}, \ \sqrt{2.414}, \ldots.$$

Yet in this case at least the objection is unfounded. For if we denote $\sqrt{1+\sqrt{2}} = x$, a simple algebraic manipulation will show that x is a solution of the equation $x^4 = 2x^2 + 1$. To this, however, a procedure similar to the algorithm of root extraction could be applied, and this would permit us to construct a set of rational approximations, which in turn would exhibit $\sqrt{1+\sqrt{2}}$ as a rational sequence, asymptotic to the irrational sequence which we first considered.

Now, strange though it may seem at first, this is a most general fact. *To any irrational sequence we can assign a rational sequence* (and usually more than one) *which is asymptotic to it.* So that the introduction of *ranked* irrationals, interesting though these latter may be from the purely formal standpoint, is quite superfluous as far as general arithmetic is concerned.

The proposition that whatever can be expressed by means of an irrational sequence is susceptible of representation by means of a sequence of rational numbers is of fundamental importance. It assigns to rational numbers a special rôle in the theory. Inasmuch as any real number can be expressed by infinite convergent rational sequences, the rational domain, reinforced by the concepts of convergence and limit, will suffice to found arithmetic, and through arithmetic the theory of functions, which is the cornerstone of modern mathematics.

But this capital fact is of just as great importance in applied mathematics. Since any rational sequence can be represented as an infinite decimal series, all computations may be systematized. By limiting himself to a certain number of decimal places, the computer may obtain a rational approximation to any irrational or transcendental problem. And what is more, the degree of accuracy of this procedure not only can be readily estimated, but even assigned in advance.

When Louis XIV was asked what was the guiding principle of his international policy, he is reported to have answered cynically: "Annexation! One can always find a clever lawyer to vindicate the act."

I am always reminded of this anecdote when I reflect on the history of the two problems: the infinite and the irrationals. The world did not wait for Weierstrass and Cantor to sanctify the procedure of substituting for an irrational number one of its irrational approximations, or, what is the same, of replacing the limit of an infinite sequence by an advanced term in the sequence. It measured its fields and erected its structures; it dug its tunnels and built its bridges; it wrought its arms and designed its machines on *rational approximations,* asking no questions as to the validity of the principle involved.

I spoke before of the approximate values which Theon and Hero gave for the square roots of integers. There are indications that the problem is of older origin and probably goes back to the early Pythagoreans. But it was Archimedes who first made the systematic applications of the principle.

Let us return again to the classical problem of squaring the circle, in which are so well reflected the various phases through which mathematics passed. As I have already mentioned, Archimede's method consisted in regarding the circumference as contained between two sets of regular polygons, one set inscribed, the other circumscribed to the circle. Starting with hexagons he kept on doubling the number until polygons of 96 sides were reached. The successive perimeters of the inscribed polygons form one sequence, and those of the circumscribed form another. If the process be continued indefinitely, the two sequences would converge towards the same limit: the length of the circumference. If the diameter of this latter be unity, the common limit is π.

What is remarkable about these sequences is that they are both irrational. Indeed, the first terms of the first are 6, 6 ($\sqrt{6} - \sqrt{2}$), the first terms of the second $4\sqrt{3}$, $24(2 - \sqrt{3})$, and the succeeding terms involve radicals of increasing complexity. These irrational sequences Archimedes, in full confidence, replaced by rational, and in this manner deduced that π was contained between the two rational numbers $3^{1}/_{7}$ and $3^{10}/_{71}$.

The same two classical problems, the radicals and the evaluation of π, gave the impetus for the development of another important infinite process: the *continued fraction*. Though some historians of mathematics maintain that these were already known to the Greeks, the first record of continued fractions which have reached us is found in a book of Bombelli dated 1572. He says, however, that "many methods of forming fractions have been given in the works of other authors; the one attacking and accusing the other without due cause, for, in my opinion, they are all looking to the same end." Judging by this, the algorithm must have been known early in the sixteenth century.

The "same end" of which Bombelli speaks is the finding of rational approximations for radicals. I shall illustrate the method on $\sqrt{2}$. This number being contained between 1 and 2, let us set $\sqrt{2} = 1 + 1/y$. From this we draw $y = 1 + \sqrt{2} = 2 + 1/y$. Continuing this way, we obtain the fraction

$$\sqrt{2} \approx 1 + \cfrac{1}{2 + \cfrac{1}{2 + \cfrac{1}{2 + \cfrac{1}{2 + \cdots}}}}$$

This is a special type of continued fraction; it is called *simple* because all the numerators are 1, and *periodic* because the denominators repeat.

If we limit ourselves to one element, two elements, three, etc., of a continued fraction, we obtain a set of rational approximations, which are called *convergents*. In the case of $\sqrt{2}$, the convergents are:

$$1, 1\frac{1}{2}, 1\frac{2}{5}, 1\frac{5}{12}, 1\frac{12}{29}, 1\frac{29}{70}, 1\frac{70}{169},....$$

Two features make continued fractions particularly valuable. *In the first place, a simple continued fraction always converges,* and, in the second place, it has an *oscillating character.* In fact, we can break up the convergents into two groups by taking the first, third, fifth, etc.; then the second, fourth, sixth, etc. In the case of $\sqrt{2}$, we obtain the two asymptotic sequences:

$$1 \qquad 1\frac{2}{5} \qquad 1\frac{12}{29} \qquad 1\frac{70}{169} \, ...$$

$$1\frac{1}{2} \qquad 1\frac{5}{12} \qquad 1\frac{29}{70} \qquad 1\frac{169}{408} \, ...$$

The first is continually increasing and has $\sqrt{2}$ for upper bound, the other is continually decreasing and has $\sqrt{2}$ for lower bound. This oscillating feature makes continued fractions invaluable for accurate approximations, for the error committed in stopping at any convergent can be readily estimated.

In the eighteenth century, Euler showed that any *quadratic irrationality can be represented by a simple periodic continued fraction*; and soon afterward Lagrange proved that the converse is true, i.e., that any such *periodic fraction represents a solution of a quadratic equation.* Thus, the periodic continued fractions play the same rôle with respect to the quadratic equation as the periodic decimal fractions with regard to the linear equation.

The procedure we established for $\sqrt{2}$ can be applied to any equation, so that any real solution of the most general equation may be represented as a continued fraction. Yet only in the case of the quadratic equation will the fraction be periodic. It would seem at first that continued fractions are peculiarly adapted to algebraic manipulations. If this were so, then we would have some criterion for distinguishing algebraic irrationals from transcendentals. Now, in so far as the algebraic origin of the fraction imposes certain limitations on the magnitude of its elements this is true, and as a matter of fact, it was this limitation that enabled Liouville to discover the *existence of nonalgebraic numbers*. But apart from this, the procedures of algebra occupy no privileged position either with regard to continued fractions, or (as far as we know) with regard to any other type of sequence. It is this remarkable "indifference" of infinite processes to algebra which is responsible for the great difficulties encountered in the theory of transcendentals.

Thus, for instance, the transcendental numbers π and e can be expressed in rather elegant fashion by continued fractions, as the reader can glean from the table at the end of this chapter.

The expansion of π into a continued fraction was discovered by Lambert in 1761, and is of great historical importance. The non-periodicity of this fraction shows conclusively that the number π is not a root of a quadratic equation with *rational coefficients*. This suggests that the quadrature of the circle cannot be achieved by straight-edge and compass alone. I say *suggests* not *proves*, for the number π may still be a root of a quadratic equation the coefficients of which involve quadratic irrationalities only, in which case the continued fraction would be non-periodic, but a straight-edge-compass construction possible.

There is a remarkable analogy between simple continued fractions and infinite decimal series. In the first place, both types of sequences are always convergent: i.e., any random law of succession of denominators of a continued fraction or digits in a decimal fraction will always represent a real number. In the second place, if the law of succession is periodic the decimal series represent a rational number, while a periodic continued fraction represents a quadratic irrationality, i.e., a number of the form $a + \sqrt{b}$, where a and b are rational. Finally, any real number can be represented either in the form of a decimal series or as a continued fraction, provided that common fractions be regarded as a special case of continued.

These properties single out these two types of infinite processes as particularly adapted for the representation of real numbers. And yet the history of infinite processes revolved about a procedure which was much more general in its scope, and which at the same time, because of its very generality and vagueness, led to a number of perplexing and paradoxical results.

No doubt the origin of this procedure was the geometrical series, which, if we are to judge from Zeno's arguments, was already known to the ancients. When we restrict ourselves to positive geometrical series, we see that the series is convergent when the ratio is less than 1, divergent otherwise. This result can be immediately generalized to *alternating* geometrical series, i.e., to the case when the ratio is negative. The alternating geometrical series, too will converge if the ratio is a proper fraction; will diverge otherwise. An interesting case, however, arises when the ratio is equal to −1. The series is then of the form

$$a - a + a - a + a - a + a - a + \dots$$

We would say today that this series diverges in spite of the fact that the sum never exceeds a. Indeed, this series can be paraphrased into the sequence:

$$a, \ o, \ a, \ o, \ a, \ o, \ a, \ o \ \dots,$$

and this has no definite limit. However, Leibnitz thought otherwise. He argued that the limits a and o are *equally probable;* and maintained that the sum approaches the mean value $1/2 \ a$ as a limit.

The tract in which Leibnitz deals with series appeared late in the seventeenth century and was among the first publications on the subject. Characteristic of this early history of series is that the question of their convergence of divergence, which today is recognized as fundamental, was in those days more or less ignored. So, for instance, it was generally believed that if the sequence generating a series is evanescent, the series is necessarily convergent. This, as we saw, holds for geometrical series, and no doubt such was the origin of this widespread error. It was not until the publication of Jacques Bernoullli's work on infinite series in 1713 that a clearer insight into the problem was gained. The occasion was the *harmonic* series

$$1 + \tfrac{1}{2} + \tfrac{1}{3} + \tfrac{1}{4} + \tfrac{1}{5} + \tfrac{1}{6} \ldots$$

The generating sequence being evanescent, it was generally held that this series converges. Bernoulli, however, gives in his book a proof, due to his brother John, of the fact that this series slowly but surely diverges.

Bernoulli's work directed attention towards the necessity of establishing criteria of convergence. The evanescence of the *general term*, i.e., of the generating sequence, is certainly a *necessary* condition, but this is generally insufficient. *Sufficient* conditions have been established by d'Alembert and Macluarin, Cauchy, Abel, and many others. I shall not dwell on this subject, which is foreign to my general purpose. However, I must state that to recognize whether a series converges or not is even today rather difficult in some cases.

There is, however, a special type of series which was in the early days a matter of considerable interest, and in which the evanescence of the general term *is* a criterion of convergences.

These are the so-called *alternating* series, typical of which is

$$1 - \tfrac{1}{2} + \tfrac{1}{3} - \tfrac{1}{4} + \tfrac{1}{5} - \tfrac{1}{6} + \tfrac{1}{7} - \ldots$$

This series converges towards the so-called *natural logarithm** of the number 2, the approximate value of which is 0.693. And yet, the series of *absolute values* is the harmonic series,

$$1 + \tfrac{1}{2} + \tfrac{1}{3} + \tfrac{1}{4} + \tfrac{1}{5} \ldots$$

which, as we stated, diverges.

Now the reason I mentioned this type of series is that it was responsible for a number of perplexing incidents. The general attitude towards series during the seventeenth and eighteenth centuries was to consider them not as special types of sequences, but as *sums* of an infinite number of terms. It was, therefore, natural to attribute to this "addition" the properties of the finite operation, i.e., associativity and commutativity. Thus, it was assumed that, the sum being independent of the arrangement of the terms, it was permissible to rearrange the terms at will.

Now, in 1848, Lejeune Dirichlet proved that such is indeed the case with a convergent series of all positive terms. If, however, the series has negative terms, then two cases may arise: if the series is *absolutely convergent,* i.e., if the series of absolute values converges, then associativity and commutativity do hold; if, however, the series is only *conditionally convergent,* i.e., the series of positive terms diverges, then these properties break down; and then, as a matter of fact, by a proper rearrangement of the terms, it is possible to deduce that the sum equals any number whatsoever.

---

*The *natural logarithm* of a number A is given by the exponent x in
$$e^x = A; \quad x = \log A$$
where *e* is the *transcendental* which was mentioned before on several occasions. The approximate value is
$$e = 2.71828$$

Small wonder, therefore, that before the days of Dirichlet many weird results were arrived at through manipulating series, particularly of the series which we called *conditionally convergent*. A historical example is the harmonic series. Let us denote by x the "sum" of its odd terms, and by y that of the even. Then we can write $y = 1 + 1/2 + 1/4 + 1/6 \ldots = 1/2 (1 + 1/2 + 1/3 + 1/4 + 1/5 + \ldots.)$ From which we carelessly draw

$$y = 1/2 (x + y) \text{ or } 1/2 \, x = 1/2 \, y \text{ or } x - y = 0.$$

We reach the fallacious conclusion that the alternating harmonic series converges toward 0, when, as a matter of fact, the series approaches the natural logarithm of 2 as a limit.

Although today such arguments would appear absurd enough, they were quite common not only in the eighteenth century but even in the early part of the nineteenth. Thus, as late as 1828, Abel in a letter to his former teacher, Holmboe, complains:

"The divergent series are the invention of the devil, and it is a shame to base on them any demonstration whatsoever. By using them, one may draw any conclusion he pleases and that is why these series have produced so many fallacies and so many paradoxes. ...I have become prodigiously attentive to all this, for with the exception of the geometrical series, there does not exist in all of mathematics a single infinite series the sum of which has been determined rigidly. In other words, the things which are the most important in mathematics are also those which have the least foundation. That most of these things are correct in spite of that is extraordinarily surprising. I am trying to find a reason for this; it is an exceedingly interesting question."

Abel's letter already breathes a new spirit. It was the dawn of a new era, the critical era in mathematics. The naïve attitude which had reigned since the beginning of the Revival of Learning was coming to an end. Tremendous conquests had been made in every field of the mathematical sciences; it became necessary now to consolidate these results into systems, and above all it was necessary to examine with care the foundations on which these systems were to rest.

To tell the story of infinite processes since Cauchy and Abel is to tell the story of modern analysis and theory of functions, and this is beyond our scope. But this sketchy presentation of the early history of infinite processes will suffice to show that the Cantor theory of irrationals was but the consummation of a long historical evolution, an evolution which began with the Pythagorean crisis, which was temporarily interrupted when the progress of all ideas was arrested, only to be resumed with the Revival of Learning.

As in the case of analysis which I presented in the last chapter, the guiding motive throughout this long period of groping was a sort of implicit faith in the *absolute nature of the unlimited.* This faith found its supreme expression in the finale: the new theories of the *continuum,* of which I am about to tell.

THE NUMBER π

$$\frac{\pi}{2} = \text{limit of product}: \frac{2\cdot 2}{1\cdot 3}\cdot\frac{4\cdot 4}{3\cdot 5}\cdot\frac{6\cdot 6}{5\cdot 7}\cdot\frac{8\cdot 8}{7\cdot 9}\cdot\frac{10\cdot 10}{9\cdot 11}\ldots\ldots$$

$$\frac{\pi}{4} = \text{limit of series}: 1-\frac{1}{3}+\frac{1}{5}-\frac{1}{7}+\frac{1}{9}-\frac{1}{11}+\frac{1}{13}-\frac{1}{15}+\frac{1}{17}-\frac{1}{19}+\frac{1}{21}\ldots\ldots$$

$$\frac{\pi}{2} = \text{limit of series}: 1+\frac{1}{3}\left(\frac{1}{2}\right)+\frac{1}{5}\left(\frac{1\cdot 3}{2\cdot 4}\right)+\frac{1}{7}\left(\frac{1\cdot 3\cdot 5}{2\cdot 4\cdot 6}\right)+\frac{1}{9}\left(\frac{1\cdot 3\cdot 5\cdot 7}{2\cdot 4\cdot 6\cdot 8}\right)+\ldots$$

$$\frac{4}{\pi} = \text{limit of fraction}: 1+\cfrac{1^2}{2+\cfrac{3^2}{2+\cfrac{5^2}{2+\cfrac{7^2}{2+\cfrac{9^2}{2+\cfrac{11^2}{2+\cdots}}}}}}$$

THE NUMBER e

$$e = \text{limit of sequence}: \left(1\tfrac{1}{2}\right)^2,\ \left(1\tfrac{1}{3}\right)^3,\ \left(1\tfrac{1}{4}\right)^4,\ \left(1\tfrac{1}{5}\right)^5,\ldots\ldots$$

$$e = \text{limit of series}: 1+\frac{1}{1!}+\frac{1}{2!}+\frac{1}{3!}+\frac{1}{4!}+\frac{1}{5!}+\frac{1}{6!}+\ldots\ldots$$

$e = \text{limit of fraction}:$

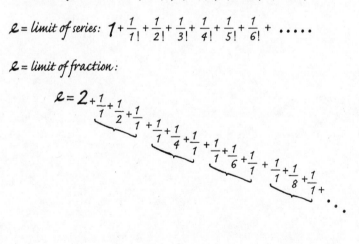

$$e = 2+\cfrac{1}{1+\cfrac{1}{2+\cfrac{1}{1+\cfrac{1}{1+\cfrac{1}{4+\cfrac{1}{1+\cfrac{1}{1+\cfrac{1}{6+\cfrac{1}{1+\cfrac{1}{1+\cfrac{1}{8+\cfrac{1}{1+\cdots}}}}}}}}}}}}$$

CHAPTER 9

Filling the Gaps

For "is" and "is-not" though with rule and line
And "up-and-down" by logic I define,
Of all that one should care to fathom, I
Was never deep in anything but—wine.

Ah, but my computations, people say,
Reduced the year to better reckoning?—Nay,
'Twas only striking from the calendar
Unborn to-morrow and dead yesterday.
                                        —Omar Khayyám

The admission of the validity of infinite process takes us out of the narrow confines of the rational domain and furnishes us with a means of attacking problems before which rational arithmetic stood powerless. It is therefore natural to inquire whether we are now in a better position to solve the old problem of establishing a perfect correspondence between the points on a line and the domain of numbers.

We know that rational arithmetic was incapable of solving this problem. But whether general arithmetic, the arithmetic of real numbers, fares better in this respect is still an open question: the points of the line which eluded rational representation, are they at all capable of arithmetical formulation? The old problem which caused the original crisis and forced a revision of the foundations of arithmetic now reappears in a new and more general form:

Can ANY real number be represented by a point on a line?
Can a real number be assigned to ANY point on a line?

If the answer is in the affirmative, then there exists a complete and reciprocal correspondence between the domain of real numbers on the one hand and the aggregate of points on the other. If such a correspondence exists, we can confidently use the intuitive language of geometry in the formulation of arithmetical analysis. and reduce these questions to number and magnitude. We see how fundamental the query is, and how much may depend on the answer!

To understand the answer that has been given to this question in modern times we must give ourselves separate accounts of the nature of the two aggregates: the domain of real numbers and the points on a line.

Of the real domain we know:

First. That it is well ordered: of any two real numbers a and b we can tell which is greater. Furthermore if a is greater than b, and b is greater than c, then also a is greater than c. In short we can by an act of the mind arrange any infinite collection of such numbers in their order of magnitude. And we can further conceive that *all* real numbers have been so arranged. This is what we mean by saying that the aggregate of real numbers is *well ordered.*

Second. That this domain has neither a first nor a last member: No matter how great a positive real number is, there is one greater; and no matter how small a negative number is, there is one still smaller. We express this by saying that the real domain extends from negative infinity to positive infinity.

Third. That among the real numbers all rational numbers are to be found. The *rational domain* is but a sub-domain in the greater real domain.

Fourth. That the aggregate of real numbers is *everywhere dense.* Between any two real numbers, no matter how small the interval, an infinite number of other real numbers may be inserted.

Does it not follow from this that the domain of real numbers is all-embracing? Well, we should be tempted to affirm this without hesitation, if it were not for our experience with the rational numbers; for let us remember that so far all we have said of the real domain could apply equally well to the rational. Yet in spite of the compact structure of the latter we found it "full of gaps." What assurance have we, indeed, that the irrationals and the transcendentals have completely filled these gaps?—that on some perhaps not so very far distant day new processes will not be discovered, which by creating new mathematical beings will expose new gaps, this time in the real domain?

To answer this question Cantor undertook to probe the fundamental difference between the rational and the real domains.

The aggregate of rational numbers, while well ordered and compact, is *imperfect.* It is imperfect because it is *not closed* with respect to infinite processes. It is not closed to infinite processes, as the very existence of irrationals shows, because there exist infinite rational sequences which, while convergent, have no rational numbers for limits. In short the aggregate of rational numbers is imperfect, because it does not contain all of its own *limiting values.*

But the aggregate of real numbers is not only well ordered and compact: it is *perfect*. It is perfect because it is *closed* to *all* infinite processes. An infinite sequence of real numbers, if convergent, represents a real number; indeed such an infinite sequence, if not rational itself, could be replaced by a rational sequence which would converge to the same limit, and this limit is by definition a real number. The aggregate of real numbers contains all its own *limiting values* and for this reason is *perfect*.

Now, not every compact aggregate is perfect, as the analysis of the rational domain shows; but every *perfect aggregate is compact*, as Cantor proved. An aggregate which is both well ordered and perfect Cantor defined as a *continuum*. The real-number domain constitutes a continuum, the *arithmetic continuum*. The domain of rational numbers, on the other hand, being imperfect, does not constitute a continuum.

And so what describes the domain of real numbers exhaustively is that it is a continuum, a continuum in the Cantor sense. Now, as we saw, the words *continuum, continuous, continuity* were used in the exact sciences from their very beginning. From time immemorial the term *continuous* has been applied to space, time and motion in the undetermined sense of something uninterrupted, something that is of the same nature in its smallest parts as it is in its entirety, something *singly connected,* in short *something continuous!* don't you know. It is one of those vague, loosely conceived notions of which intuition perceives the sense; and yet any attempt to formulate it in a precise definition invariably ends in an impatient:"Well, you know what I mean!"

The prototype of all ideas which meet these specifications is the line, and particularly the straight line, which in our mind is endowed with this continuity *par excellence*. So that if we are to have a complete and reciprocal correspondence between the line and the real domain we must make sure that there is no flagrant contradiction between this intuitive idea of continuity we ascribe to the line and the precise, scientifically formulated continuity of real numbers as defined by Cantor.

If, without venturing on a precise formulation of our intuitive idea of continuity, I should try to describe roughly what I mean by *continuous,* I should be thinking aloud as follows:

> "Time is the essence of all things. Mother Nature *makes* no jumps, because Father Time *knows* no jumps. Time cannot be conceivably interrupted, that is why there is nothing spontaneous in nature. Time flows on and in its flow it carries all things conceivable."

And so when we attempt to describe the continuity of any phenomenon, we find ourselves invariably, even if unconsciously, invoking the continuity of time. The line appears to us as the prototype of all things continuous because we conceive it as generated by a continuous passage, because to our minds it is but a concrete representation of the Stream of Time, frozen as it were.*

So too, is it with other phenomena. The mind shrinks before the spontaneous; that is why our scientific theories cling so desperately to evolution. Be it a cosmogony, a theory of life, or a sociological hypothesis, everywhere we find this horror of the cataclysm. At any cost we refuse to recognize that catastrophe and revolution, spontaneous generation and accidental discovery, may have been dominant factors in the history of the universe or of the race.

And just as evolution gives us a smooth picture of our past, the doctrine of causality, by linking all phenomena into one continuous chain, safeguards our future against all spontaneous disturbances and protects us against the horror of chaos. These vague ideas of continuity and causality are so closely associated

*See Appendix 20.

that one is constantly invoked to support the other. And no wonder: our belief in the continuity of the universe and our faith in the causal connection between its events are but two aspects of this primitive intuition that we call *time*. And so on the one hand there is the conviction that *Natura non facit saltus*, and on the other hand arises the illusion: *post hoc, ergo propter hoc.*

Herein I see the genesis of the conflict between geometrical intuition, from which our physical concepts derive, and the logic of arithmetic. The harmony of the universe knows only one musical form—the *legato;* while the symphony of number knows only its opposite—the *staccato.* All attempts to reconcile this discrepancy are based on the hope that an accelerated *staccato* may appear to our senses as a *legato.* Yet our intellect will always brand such attempts as deceptions and reject such theories as an insult, as a metaphysics that purports to explain away a concept by resolving it into its opposite.

But these protests are in vain. To bridge the chasm between the continuity of our concept of time and the inherent discontinuity of the number structure, man had to invoke once more that power of his mind "which knows itself capable of conceiving the indefinite repetition of the same act when once this act is possible." This was the historic rôle of the infinite; this is why through the ages the problems of the continuum and of the infinite were but the two horns of a single dilemma. This long process of adaptation has now culminated in the Cantor theory, in which any number is conceived as the *goal* of an infinite succession of jumps, and the continuum is regarded as comprising not only all possible *resting stations* but all possible goals as well. It is a staccato theory *par excellence;* and yet it does not escape the tyranny of time. It merely adapts itself to this tyranny by complacently regarding the flowing stream of duration as an infinite succession of pulsations of furiously accelerated tempo.

The mind rebels against this tyranny, the mind demands a theory of number free from the extraneous influences of geometry or mechanics. So history was to witness one more gesture: this took the form of a new theory of irrationals which bears the name of its author, Richard Dedekind.

The essence of the Dedekind concept is contained in the following passage taken from his epoch-making essay "Continuity and Irrational Numbers" which appeared in 1872, ten years before the Cantor essays on the same subject were published:

"The straight line is infinitely richer in point-individuals than the domain of rational numbers is in number-individuals. ...

"If then we attempt to follow up arithmetically the phenomena which govern the straight line, we find the domain of rational numbers inadequate. It becomes absolutely necessary to improve this instrument by the creation of new numbers, if the number domain is to possess the same completeness, or, as we may as well say now, the same continuity, as the straight line. ...

"The comparison of the domain of rational numbers with a straight line has led to the recognition of the existence of gaps, of a certain incompleteness or discontinuity, in the former; while we ascribe to the straight line completeness, absence of gaps, or continuity. Wherein then does this continuity consist? Everything must depend on the answer to this question, and only through it shall we obtain a scientific basis for the investigation of all continuous domains. By vague remarks upon the unbroken connection in the smallest part, nothing, obviously, is gained; the problem is to indicate a precise characteristic of continuity that can serve as a basis for valid deduction. For a long time I pondered over this in vain, but finally I found what I was seeking. This discovery will perhaps be differently estimated by different people; the majority may find its substance very commonplace. It consists in the following. In the preceding section attention was called to the fact that every point of the straight line produces a separation of it into two portions such that every point of one portion lies to the left of every point of the other. I find the essence of continuity in the converse, i.e., in the following principle:

"If all points of a straight line fall into two classes, so that every point of the first class lies to the left of every point of the second class, then there exists one and only one point which produces this division of all points into two classes, this severing of the straight line into two portions.

"As already said, I think I shall not err in assuming that every one will at once grant the truth of this statement; moreover, the majority of my readers will be very much disappointed to learn that by this commonplace remark the secret of continuity is to be revealed. To this I may say that I am glad that every one finds the above principle so obvious and so in harmony with his own ideas of a line; for I am utterly unable to adduce any proof of its correctness, nor has any one else the power. The assumption of this property of the line is nothing else than an axiom by which we attribute to the line its continuity, by which we define its continuity. Granted that space has a real existence, it is *not* necessary that it be continuous; many of its properties would remain the same if it was discontinuous. And if we knew for certain that space was discontinuous, there would be nothing to prevent us, in case we so desired, from filling up its gaps in thought, and thus making it continuous; this filling-up would consist in the creating of new point-individuals, and this would have to be effected in accordance with the above principle."

Let us analyze the Dedekind principle at work. Like Cantor, Dedekind takes for his point of departure the domain of rational numbers. However, instead of identifying the real number with a convergent sequence of rationals, he views the real number as generated by the power of the mind to classify rational numbers.

This special classifying scheme he calls *schnitt,* a term which has been variously translated as the *Dedekind cut, split, section,* and *partition.* I choose the last.

This partition is the exact counterpart of the concept which Dedekind used in defining the continuity of a line. Just as any point on the line severs the line into two contiguous, non-overlapping regions, so does every real number constitute a means for dividing all rational numbers into two classes which have no element in common, but which together exhaust the whole domain of rational numbers.

Conversely, any equation, any scheme of classification, any process, which is capable of effecting such a split in the domain of rational numbers is *ipso facto* identified with a number, it is by definition a real number, an element of the new domain.

The rational numbers are a part of this vast domain because each one individually can be regarded as such a classifying scheme. Indeed, with respect to any given rational number, say 2, all rational numbers can be divided into two classes: those that are less than 2 or equal to it go into the *lower class;* those that are greater than 2, into the *upper class.* The two classes have no elements in common and together they exhaust the whole aggregate of rational numbers. The rational number 2 may be regarded as a *partition* and is therefore a real number.

But obviously the potentialities of this far-reaching principle cannot be exhausted by such trivial partitions. Nothing, for instance, can prevent us from partitioning all rational numbers into those whose square is less than or equal to a given rational number, say 2, and those whose square is greater than 2. These two classes are mutually exclusive, as was the case in the previous example; and, as before, the two classes taken together exhaust all rationals. This partition too defines a real number which we identify with our old friend, $\sqrt{2}$.

On the other hand, while both rational and irrational numbers may be represented by partitions, the choice of rationals for basis is not without consequence, for there is an essential difference between rational and irrational partitions. The rational partition is itself a part of the lower class: it is as if a *politician* had split a party and joined the left wing. But the irrational partition is completely *ex parte:* it is as if an *issue* had split the party, the issue being a part neither of the left nor of the right wing. And so here the irrational that caused the partition belongs neither to the lower nor yet to the upper class. In other words: in the rational case, the lower class has a greatest element and the upper no least; in the irrational case, the lower class does not have a greatest element, nor yet does the upper class have a least.

According to the Dedekind theory this is the only feature which distinguishes the two types of number: it is characteristic of a rational number *to belong to one* of the classes, and it is just as characteristic of the irrational *to belong to neither.*

To prove that the Dedekind partitions are *bona fide* numbers it must be shown that they satisfy all the stipulations of the principle of permanence. What I said in the preceding section proves that the first stipulation is satisfied. That this is also true of the others can be proved with utmost simplicity and perfect rigor. The criteria of rank; the definition of operations; the proof of the associative, the commutative, and the distributive properties of these operations—all are the exact counterparts of the corresponding propositions in the Cantor theory, and I shall not bore the reader with the details.

The fundamental theorem in the Cantor theory—that the domain of real numbers is closed with respect to infinite processes—also has its counterpart in the Dedekind theory: having defined the domain of reals, it is natural to inquire again whether a subsequent application of the principle would not still further extend the domain. In other words, let a partition be now effected which would split all the *real* numbers into two classes. Could such a partition create a new type of magnitude, not found among the *real* numbers? The answer is *no;* any such partition can be brought about by a partition in the rational domain. The aggregate of all partitions of the rational domain is *closed.*

The complete equivalence of the two theories of the arithmetic continuum has been recognized by the authors themselves, and their rivalry, if there ever was a rivalry, is today only an historical incident. To any partition in the rational domain corresponds a limiting value of an infinite sequence; and conversely a limiting value of any infinite sequence can be used as an agent for the partitioning of the rational domain. All conceivable partitions on the one hand, all the limiting values of rational sequences on the other, are identical, and are just two descriptions of the same aggregate, the arithmetic continuum.

From the metaphysical standpoint this seems very puzzling indeed. As I have said before, the Cantor theory is the consummation of a long historical process; the Dedekind partition is a bold and original conception. The Cantor theory uses the infinite process to generate the number domain; whereas nowhere in the definition of the real number does Dedekind use the word *infinite* explicitly, or such words as *tend, grow beyond measure, converge, limit, less than any assignable quantity,* or other substitutes. Again the Cantor theory is frankly *dynamic:* the limiting value is being generated in a manner which strongly resembles the motion of a point attracted to a center. The Dedekind theory is essentially *static,* no other principle being utilized than this power of the mind to classify elements along a definite scheme. So it seems at a first glance that here we have finally achieved a complete emancipation of the number concept from the yoke of the time intuition which long association with geometry and mechanics had imposed upon it.

And yet the very equivalence of two theories so opposite in their points of departure, and so different in their modes of attack, shows that things are not so happy with the Dedekind principle as may first appear. And indeed a further analysis of the Dedekind procedure discloses that the infinite here is implied, even if not explicitly used. The principle of partitioning if effected upon a *finite* set of rational numbers would lead to trivialities, which would at once betray its futility. Furthermore any practical application of the principle towards the determination of an irrational number necessitates the employment of a machinery analogous to the infinte sequence of Cantor.

And the same is true of the relation of the theory to the time intuition. The axiom of Dedekind—"if all points of a straight line fall into two classes, such that every point of the first class lies to the left of any point of the second class, then there exists one and only one point which produces this division of all points into two classes, this severing of the straight line into two portions"—this axiom is just a skilful paraphrase of the fundamental property we attribute to time. Our intuition permits us, by an act of the mind, to sever *all time* into the two classes, *the past* and *the future,* which are mutually exclusive and yet together comprise all of time, *eternity.* The *now* is the partition which separates all the past from all the future; any instant of the past was once a *now,* any instant of the future will be a *now* anon, and so any instant may itself act as such a partition. To be sure, of the past we know only disparate instants, yet, by an act of the mind we fill out the gaps; we conceive that between any two instants— no matter how closely these may be associated in our memory— there were other instants, and we postulate the same compactness for the future. This is what we mean by the flow of time.

Furthermore, paradoxical though this may seem, *the present is truly irrational* in the Dedekind sense of the word. For while it acts as partition it is neither a part of the past nor a part of the future. Indeed, in an arithmetic based on pure time, if such an arithmetic was at all possible, it is the irrational which would be taken as a matter of course, while all the painstaking efforts of our logic would be directed toward establishing the existence of rational numbers.

Finally, when Dedekind says that "if we knew for certain that space was discontinuous, there would be nothing to prevent us, in case we so desired, from filling up its gaps in thought and thus making it continuous," he states a *post factum*. This filling-out process was accomplished ages ago, and we shall never discover any gaps in space for the simple reason that we cannot conceive of any gaps in time.

Yet in spite of the fact that neither Cantor nor Dedekind succeeded in emancipating the continuous from the intuition of time, the age-old conflict between our notions of continuity and the scientific concept of number ended in a decisive victory for the latter. This victory was brought about by the necessity of vindicating, of legitimizing, as it were, a procedure which ever since the days of Fermat and Descartes had been an indispensable tool of analysis. The history of this discipline, *analytic geometry,* forms a part of the next chapter. It will suffice for my purposes here to state that this discipline which was born of the endeavors to subject problems of geometry to arithmetical analysis, ended by becoming the vehicle through which the abstract properties of number are transmitted to the mind. It furnished analysis with a rich, picturesque language and directed it into channels of generalization hitherto unthought of.

Now, the tacit assumption on which analytic geometry operated was that it was possible to represent the points on a line, and therefore points in a plane and in space, by means of numbers. This assumption is of course equivalent to the assertion that a perfect correspondence can be established between the points on a line and the real numbers. The great success of analytic geometry, the fact that it so admirable served the purposes of both analysis and geometry, gave this assumption an irresistible pragmatic force. It was essential to include this principle in the general logical structure of mathematics. But how?

Under such circumstances mathematics proceeds by *fiat*. It bridges the chasm between intuition and reason by a convenient postulate. This postulate ousts the intuitive notion and substitutes for it a logically consistent concept. The very vagueness of all intuition renders such a substitution not only plausible, but highly acceptable.

And so it happened here. On the one hand, there was the logically consistent concept of the real number and its aggregate, the arithmetic continuum; on the other hand, the vague notions of the point and its aggregate, the linear continuum. All that was necessary was to declare the identity of the two, or, what amounted to the same thing to assert that:

It is possible to assign to any point on a line a unique real number, and, conversely, any real number can be represented in a unique manner by a point on the line.

This is the famous *Dedekind-Cantor* axiom.

This proposition, by sanctifying the tacit assumption on which analytic geometry had operated for over two hundred years, became the fundamental axiom of this discipline. This axiom, like so many others, is really a disguised definition: it defines a new mathematical being, the *arithmetical line.* Henceforth the line—and consequently the plane, and space—ceases to be an intuitive notion and is reduced to being *a mere carrier of numbers.*

And so this axiom is tantamout to an arithmetization of geometry. It means the emancipation of analysis from geometrical intuition, to which it owed its birth and growth. It means more: it is a bold declaration that henceforth analysis proposes to assume control over geometry and mechanics, and through them to control those other phases of our cognition, which are even nearer to the crude reality of our senses.

The age-long struggle to create an arithmetic in the image of that reality had failed because of the vagueness of that reality. So arithmetic created a new reality in its own image. The infinite process succeeded where the rational number had failed.

Numeri mundum regnant.

CHAPTER 10

The Domain of Number

"To Archimedes came a youth eager for knowledge.
Teach me, O master, he said, that art divine
Which has rendered so noble a service to the lore of the
heavens,
And back of Uranus yet another planet revealed.
Truly, the sage replied, this art is divine as thou sayest,
But divine it was ere it ever the Cosmos explored,
Ere noble service it rendered the lore of the heavens,
And back of Uranus yet another planet revealed.
What in the Cosmos thou seest is but the reflection of God,
The God that reigns in Olympus is Number Eternal."

—K. G. J. Jacobi

By trying and erring, by groping and stumbling—so progressed our knowledge. Hampered and yet spurred by a hard struggle for existence, a plaything of his environment and a slave to the traditions of his time, man was guided in this progress not by logic but by intuition and the stored-up experience of his race. This applies to all things human, and I have made painstaking efforts to show that mathematics is no exception.

And yet who knows but that the habit of systematic exposition, acquired by years of teaching, has caused me to transgress unwittingly? Now, the evolution of number, when

187

presented in broad outline, does seem to possess a certain logical continuity. But a broad outline is usually a crude outline: it teaches little of true significance. The irregularities of a curve apprise us more of its true nature than does its shape; similarly, the irregularities in the development of any human endeavor bring out more clearly the underlying factors than do those features that it has in common with similar endeavors.

The systematic exposition of a textbook in mathematics is based on logical continuity and not on historical sequence; but the standard high-school course, or even the college course in mathematics fails to mention this fact, and therefore leaves the student under the impression that the historical evolution of number proceeded in the order in which the chapters of the textbook were written. This impression is largely responsible for the widespread opinion that mathematics has no human element. For here, it seems, is a structure that was erected without a scaffold: it simply rose in its frozen majesty, layer by layer! Its architecture is faultless because it is founded on pure reason, and its walls are impregnable because they were reared without blunder, error or even hesitancy, for here human intuition had no part! In short the structure of mathematics appears to the layman as erected not by the erring mind of man but by the infallible spirit of God.

The history of mathematics reveals the fallacy of such a notion. It shows that the progress of mathematics has been most erratic, and that intuition has played a predominant rôle in it. Distant outposts were acquired before the intermediate territory had been explored, often even before the explorers were aware that there was an intermediate territory. It was the function of intuition to create new forms; it was the acknowledged right of logic to accept or reject these forms, *in whose birth it had no part.* But the decisions of the judge were slow in coming, and in the meantime the children had to live, so while waiting for logic to sanctify their existence, they throve and multiplied.

The evolution of the complex number concept, this weird- est chapter in the history of mathematics, bears all the marks of such a development. Did the science of number wait for Weierstrass and Cantor and Dedekind to establish the real num- ber on a logical foundation, before venturing on new conquests? No; taking the legitimacy of the real number for granted, it pro- ceeded to explore another mystic corner of its world, and emerged from this expedition the master of a new domain of unprecedented magnitude and promise.

We want to trace the new concept to its source. So I resume the study of the problem in algebra which was the source of the real number. We were interrupted in this study by our long excursion into the infinite. We now return to it with a number concept vastly enriched, strongly reinforced by our new weapon: the infinite process. Instead of the rational aggregate, we have now at our disposal the arithmetic continuum; and in addition to the rational processes of finite algebra we are now aided by the powerful machinery of analysis. Surely, we should now be in a position to attack with assurance the general equation of algebra!

Well, the reader who remembers his elementary algebra knows that such is not the case: the real number too is inade- quate for the solution of all equations of algebra. And to prove this it is not necessary to construct intricate equations of higher degree. It is sufficient to consider one of the simplest equations of its kind, the quadratic: $x^2 + 1 = 0$.

It defines no Dedekind partition, nor can we construct a Cantor sequence the square of which would converge towards -1

as a limit. In the twelfth century the Brahmin Bhaskara expressed this in the simple and effective statement:

> "The square of a positive number, as also that of a negative number, is positive; and the square root of a positive number is two-fold, positive and negative; there is no square root of a negative number, for a negative number is not a square."

The impulse to write $x = \sqrt{-1}$ as a solution of this equation was curbed by the knowledge that such an expression had no concrete meaning. The Hindu mathematicians resisted this temptation and so did the Arabs. The glory of having discovered the *imaginary* goes to the Italians of the Renaissance. Cardan in 1545 was the first to dare to denote the meaningless by a symbol. In discoursing on the impossibility of splitting the number 10 into two parts, the product of which was 40, he showed that the formal solution would lead to the impossible expressions: $5 + \sqrt{-15}$ and $5 - \sqrt{-15}$.

But, as happened in the case of the negative numbers, so here too the mere writing down of the impossible gave it a symbolic existence. It is true that it was written with the reservation that it was *meaningless, sophisticated, impossible, fictitious, mystic, imaginary.* Yet, there is a great deal in a name, even if it be but a nickname or a term of abuse.

Strangely enough it was not the quadratic but the cubic equation which gave the impetus towards handling these mystic beings as *bona fide* numbers. This happened in the following way:

The cubic equation $x^3 + ax + b = 0$ has at least one real solution, and may have three. Now in the case when only one solution is real, Scipio del Ferro, Tartaglia and Cardan developed a procedure which is summed up in the so-called Cardan formula. This formula, however, breaks down when all three roots are real, for in this case the radicals entering into the formula represent *imaginary* numbers.

Thus consider the historic equation $x^3 = 15x + 4$, treated by Bombelli in his algebra published in 1572. This equation has three real solutions; namely, 4, $(-2 + \sqrt{3})$, and $(-2 - \sqrt{3})$. Yet an application of the Cardan formula leads to the purely illusory result

$$x = \sqrt[3]{2 + \sqrt{-121}} + \sqrt[3]{2 - \sqrt{-121}}.$$

Now it occurred to Bombelli that perhaps the two radicals represent expressions of the type $p + \sqrt{-q}$ and $p - \sqrt{-q}$, expressions which today we would call *conjugate complex*. If such were the case, and if the addition of such beings could be performed according to the usual rules, then the sum of two such "sophisticated" magnitudes may give a real number, and perhaps even one of the actual solutions of the equation, which Bombelli knew to be 4. But let me leave the word to Bombelli himself:

"It was a wild thought, in the judgment of many; and I too was for a long time of the same opinion. The whole matter seemed to rest on sophistry rather than on truth. Yet I sought so long, until I actually proved this to be the case."

Indeed Bombelli showed that the two cubic radicals resolve into $2 + \sqrt{-1}$ and $2 - \sqrt{-1}$, the sum of which is 4.

Impossibly these beings were, yes! But not altogether useless, since they could serve as the instrument for solving real problems. So Bombelli, encouraged by his success, proceeded to develop rules for operations on these complex beings.

Today we have simplified Bombelli's notation by adopting the symbol i for $\sqrt{-1}$. Any complex being is of the type $a + ib$. With this notation the solution of Bombelli's equation is $x = \sqrt[3]{2 + 11i} + \sqrt[3]{2 - 11i} = (2 + i) + (2 - i) = 4$.

We call these Bombelli beings *complex numbers,* and to justify the name *number* we prove that they satisfy all the stipulations of the principle of permanence. Bombelli knew nothing of this principle; he was guided solely by his mathematical conscience, for which intuition is another name. Yet apart from the notation, the gifted Italian had all the rules practically in the form in which they are taught today.

The first stipulation is satisfied because the complex $a + ib$ comprises the reals as a sub-domain ($b = 0$). The criterion of rank consists in stipulating that $a + ib$ and $c + id$ are equal if $a = c$ and $b = d$, unequal otherwise. As to the criterion of greater or of less, they are not so straightforward. However, the difficulties encountered are not serious enough to deserve special mention.

The sum of two complex numbers is a complex number which is obtained by separately adding the real and the imaginary parts; similarly for the difference. The product of two or more complex numbers is obtained by multiplying these individual numbers according to the ordinary rules of algebra and replacing everywhere the powers of i according to the schema

$i = \sqrt{-1}$	$i^5 = i$	$i^9 = i$	
$i^2 = -1$	$i^6 = -1$	$i^{10} = -1$	
$i^3 = -i$	$i^7 = -i$	$i^{11} = -i$	etc., etc.
$i^4 = 1$	$i^8 = 1$	$i^{12} = 1$	

The Bombelli operations therefore are commutative, associative, and distributive. All the stipulations of the principle are satisfied. Thus is created the *complex number domain,* which supersedes the real domain in the same way as the latter has superseded the rational.

As a corollary it follows that any set of rational operations performed on complex numbers leads to complex numbers. In other words, the domain of the complex is *closed* with respect to rational operations.

Is it closed to the infinite processes of analysis as well? Or in other words, can we extend the notion of infinite sequence, of convergence and limit, to embrace the complex numbers too? An affirmative answer to this query was given in the nineteenth century by Gauss, Abel, Cauchy, and Weierstrass, and this fundamental fact forms the foundation of the modern theory of functions.

Even in the eighteenth century the complex number had already begun to lose its purely algebraic character. The famous identity discovered by de Moivre showed the rôle the complex number played in trigonometry, while Euler amplified de Moivre's formula by bringing in the transcendental *e*. Although this is somewhat beyond my scope I must, for the sake of completeness, mention this striking identity of Euler,

$$e^{i\pi} + 1 = 0,$$

which was considered by some of his metaphysically inclined contemporaries as of mystic significance. Indeed, it contains the most important symbols of modern mathematics and was regarded as a sort of *mystic union,* in which arithmetic was represented by 0 and 1, algebra by the symbol *i*, geometry by π, and analysis by the transcendental *e*.

It is natural to inquire whether the instrument created by the adjunction of the complex numbers is adequate for the solution of the fundamental problem of algebra: determining a root of the most general equation.

Already Bombelli knew that by means of complex numbers the quadratic and cubic equation can be completely solved; in other words, that the most general equation of the second and third degree must possess at least one root which may be a real or a complex number. This followed from the fact that these equations lead to formal solutions in terms of quadratic and cubic surds. To be sure, the latter may involve complex numbers, but such radicals may themselves be resolved into the form $a + ib$.

Since the Ferrari method establishes a similar procedure for the equations of the fourth degree, these too have solutions which can be expressed as complex numbers, the case of a real solution being a particular instance.

These facts were known in the seventeenth century. It was also known that imaginary roots of an algebraic equation with real coeffecients must come in pairs, i.e., that if $a + ib$ was a solution of such an equation, the conjugate $a - ib$ was also one. From this it followed that an equation of odd degree must possess at least one *real root*.

Now in 1631 the Englishman Thomas Harriot came out with the ingenious idea of putting any equation into the form of a polynomial equal to 0; a far-reaching thought, for it led Harriot to the theorem (which today we call the factor theorem) that if a be a root of an algebraic equation, then $x - a$ is a factor in the corresponding polynomial. This fundamental fact reduced the solution of any equation to a problem in factoring, and showed conclusively that if it could be proved that any equation has a root, real or complex, it would be *ipso facto* established that the equation has as many roots as its degree indicates; with the reservation, of course, that *every root be counted as many times as the corresponding factor enters in the polynomial.*

It was surmised by Girard early in the seventeenth century that what was true for equations of the first four degrees was generally true; and in the middle of the eighteenth century d'Alembert formulated it in the statement that *any algebraic equation must possess at least one solution real or complex.* He, however, was unable to prove this assertion rigorously, and in spite of the efforts of many who followed him it remained a postulate for another fifty years.

This assertion recalls the other statement: any equation can be solved by means of radicals. This, too, we saw, was considered obvious by many mathematicians even in the days of Lagrange. Yet the comparison is unfair: here the generalization was of the type called incomplete induction, and the falsity of the proposition only brought out in relief the danger of this method. Entirely different is the intuition which led to d'Alembert's postulate.

This intuition is reflected in all the proofs of this *fundamental theorem of algebra* that have been given since the days of d'Alembert; namely, d'Alembert's, Euler's, and Lagrange's insufficient demonstrations; the proofs which Argand gave for it in 1806 and 1816; the four proofs by which the great Gauss established the proposition; and all subsequent improvements on these latter.

Different though these proofs are in principle, they all possess one feature in common. Somewhere, somehow,—sometimes openly, sometimes implicitly—the idea of continuity is introduced, an idea which is foreign to algebra, an idea which belongs to the realm of analysis.

Let me explain this by a simple example. If we set $Z = z^2 + 1$ and $z = x + iy$, we obtain upon substitution $Z = (x^2 - y^2 + 1) + i(2xy)$. Now when x and y vary in a continuous manner and assume all possible values between $-\infty$ and $+\infty$, the expressions within the parentheses will also assume all possible values within the same range. To *prove* this in the general case and in full rigor is a matter of considerable difficulty; towards this end were directed the unsuccessful efforts of d'Alembert and the genius of Gauss. But to *conceive* that it is so is another matter; here is where the *intuition of continuity* did its work. For certain values of the variables x and y the polynomials are positive, for others they are negative. The variation being continuous, there exist intermediate values of x and y, and in fact an infinity of such, which will render the first polynomial zero; and there exists another range of values for which the second polynomial will vanish. These two ranges will have *some pairs in common*. If a and b are such a pair, then $a + ib$ is a root of the equation $Z = 0$. This is what mathematical intuition suggested, and this is what d'Alembert tried to prove. Gauss succeeded where d'Alembert failed, and yet the fact that his first proof of this fundamental theorem of algebra depended on considerations of analysis rankled in his mind. So sixteen years later he produced another proof. He showed that any equation of even order can by purely algebraic means be reduced to an equation of odd degree. The fundamental theorem would then be established if it could be demonstrated that an equation of an odd degree must possess at least one real root. But unfortunately this latter proposition cannot be demonstrated, either, without bringing in considerations foreign to pure algebra.

The very fact that the proof of the fundamental theorem of algebra implies processes foreign to algebra suggests that the theorem may be of a more general scope. And such is indeed the case. The property of possessing a solution within the domain of complex numbers is not at all the monopoly of algebraic equations. Such equations as $e^z + z = 0$, for instance, and many others of a transcendental form, also admit complex solutions.

The polynomials constitute only an extremely small portion of a class of functions which Weierstrass named *entire*. Like the polynomials these functions will, for proper values of the variable, assume any complex value assigned in advance, and in particular the value zero. And to this class belong the most important transcendental expressions, such as the sine, the cosine and the exponential function. From the standpoint of the theory of functions, the *entire* functions are an immediate extension of polynomials.

Such is the basis of the theory of functions of a complex variable established by Cauchy, Weierstrass, and Riemann, a theory which was destined to become the dominant factor in the development of mathematics in the nineteenth century.

But let me return to my narrative.

In 1770, appeared Euler's Algebra, in which a great number of applications of complex magnitudes was given. Yet we read there:

"All such expressions as $\sqrt{-1}, \sqrt{-2}$, etc., are consequently impossible or imaginary numbers, since they represent roots of negative quantities; and of such numbers we may truly assert that they are neither nothing, nor greater than nothing, nor less than nothing, which necessarily constitutes them imaginary or impossible."

In 1831, Gauss wrote:

"Our general arithmetic, so far surpassing in extent the geometry of the ancients, is entirely the creation of modern times. Starting originally from the notion of absolute integers it has gradually enlarged its domain. To integers have been added fractions, to rational quantities the irrational, to positive the negative, and to the real the imaginary. This advance, however, had always been made at first with timorous and hesitating steps. The early algebraists called the negative roots of equations false roots, and this is indeed the case, when the problem to which they relate has been stated in such a form that the character of the quantity sought allows of no opposite. But just as in general arithmetic no one would hesitate to admit fractions, although there are so many countable things where a fraction has no meaning, so we ought not deny to negative numbers the rights accorded to positive, simply because innumerable things admit of no opposite. The reality of negative numbers is sufficiently justified since in innumerable other cases they find an adequate interpretation. This has long been admitted, but the imaginary quantities,—formerly and occasionally now improperly called impossible, as opposed to real quantities,—are still rather tolerated than full naturalized; they appear more like an empty play upon symbols, to which a thinkable substratum is unhesitatingly denied even by those who would not depreciate the rich contribution which this play upon symbols has made to the treasure of the relations of real quantities.

"The author has for many years considered this highly important part of mathematics from a different point of view, where just as objective an existence can be assigned to imaginary as to negative quantities, but hitherto he has lacked the opportunity to publish his views."

What has happened in the sixty years which separate the two statements to bring about such a radical change of front? Gauss answers this in his own words: "An objective existence can be assigned to these imaginary beings." In other words, a concrete interpretation similar to that which identifies negative numbers with a change in sense.

To understand this interpretation thoroughly we must take a glance back into the seventeenth century and survey in retrospect a discipline to which I repeatedly referred in the preceding chapters: analytic geometry.

When we hear about the profound changes which science

has wrought in our lives we think of physics and chemistry. We find palpable evidence of this tremendous upheaval in the mechanical inventions which have revolutionized industry and transportation. The applications of electricity have reduced the drudgery of domestic duties and developed communication between people to undreamed-of extents. The achievements of chemistry have permitted us to convert heretofore useless materials into sources of subsistence, comfort and pleasure. All this has taught man to respect and marvel at the accomplishment of these sciences.

Less evident, because more diffused, are the benefits which mathematics has conferred upon us. It is true, we know, that mathematics has played its rôle in the theories which made these inventions possible, as well as in the design of these inventions. Yet this is a matter for specialists. While in his daily life man may profit by the knowledge of the elements of which water is composed or of the difference between short and long waves, the study of geometry or calculus will contribute little to his happiness.

There are, however, among the rich achievements of mathematics some which even in this direct sense can be considered as useful inventions, because they have penetrated into the daily life of the people. To these belong our positional numeration, which made calculation accessible to the average mind; of such immediate usefulness is also the symbolism of algebra, particularly the *logistica speciosa* of Vieta, which put at the disposal of the many contracted forms of general relations, heretofore comprehensible only to the few. To this category also belongs the great invention which Descartes gave the world, the *analytical diagram*, which gives at a glance a *graphical* picture of the *law* governing a phenomenon, or of the *correlation* which exists between dependent events, or of the *changes* which a situation undergoes in the course of time.

It is a remarkable fact that the mathematical inventions which have proved to be most accessible to the masses are also those which exercised the greatest influence on the development of pure mathematics. The principle of position gave us *zero* without which the concept of negative number would not have developed; it gave the possibility of standardizing equations and made the factor theorem possible. The literal notation changed mathematics from the study of the particular to that of the general, and by symbolizing the impossible it prepared the road for the generalized number concept.

Finally the invention of Descartes not only created the important discipline of analytic geometry, but it gave Newton, Leibnitz, Euler, and the Bernoullis that weapon for the lack of which Archimedes and later Fermat had to leave inarticulate their profound and far-reaching thoughts.

"Proles sine matre creata."

"Children not born of a mother." In these words did the geometer Chasles characterize the achievement of Descartes. With equal injustice to what preceded them, could this have been said of the principle of position and of the literal notation! The former we traced to the empty column of the counting board, and the latter, we saw, was but a development of a rhetorical symbolism practiced by mathematicians and near-mathematicians since time immemorial.

Similarly the great Cartesian invention had its roots in those famous problems of antiquity which originated in the days of Plato. In endeavoring to solve the problems of the trisection of an angle, of the duplication of the cube and of the squaring of the circle, the ruler and compass having failed them, the Greek geometers sought new curves. They stumbled on the *conic sections,* i.e., the curves along which a general plane may cut a

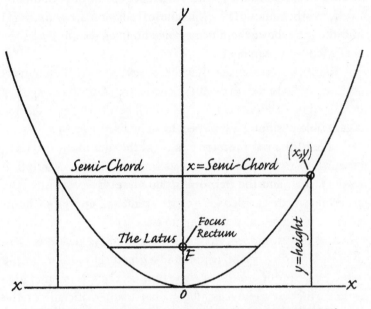

DISGUISED ANALYTICAL GEOMETRY: APPOLONIUS'
(OF PEIGA) TREATMENT OF THE PARABOLA

circular cone: the *ellipse,* the *hyperbola,* and the *parabola.* Their elegant properties so fascinated the Greek geometers that soon these curves began to be studied for their own sake. The great Apollonius wrote a treatise on them, in which he described and demonstrated the most important properties which these curves possess.

There we find the nucleus of the method which Descartes later erected into a principle. Thus Apollonius referred the parabola to its axis and principal tangent, and showed that the semichord was the mean proportional between the *latus rectum* and the height of the segment. Today we express this relation by the Cartesian equation $x^2 = Ly$, calling the height the *ordinate* (y) and the semichord the *abscissa* (x); the *latus rectum* being the coefficient of y, namely L.

Significant is the fact that the Greeks named these curves and many others which they discovered *loci;* that is, they described these curves as the *places* of all points which had some measurable position with respect to some fixed reference system. Thus the ellipse was the *locus* of a point the sum of the distances of which from two fixed points was constant. Such a description was in fact a *rhetorical equation* of the curve, for it furnished the criterion which enabled one to ascertain for any given point whether it did or did not belong to the curve.

And it was, indeed, in this sense that these relations were used by Omar Khayyám, who found a graphical solution for the cubic equation by means of two conic sections. Such methods were further developed by the Italian mathematicians of the Renaissance and by Vieta. In fact it was problems of this nature that led the latter to develop his *logistica speciosa*.

And last, but by no means least, consider this passage from an essay written by Fermat in 1629, but not published until forty years later, thirty years after the appearance of Descartes's *Géométrie:*

> "Whenever two unknown magnitudes enter into a final equation, we have a *locus,* the extremity of one of the unknown magnitudes describing a straight line or a curve. The straight line is simple and unique, the classes of curves are indefinitely many: circle, hyperbola, parabola, ellipse, etc …

"It is desirable in order to aid the concept of an equation to let the two unknown magnitudes form an angle which we would assume to be a right angle."

No, Cartesian geometry was anything but a motherless child. At the risk of seeming facetious I shall say that not only had the Descartes conception a mother—the geometry of the Greeks—but it had a twin-brother as well. Indeed, even a superficial study of Descartes's *Géométrie* and Fermat's *Introduction* discloses that we have before us one of those twin-phenomena in which the history of mathematics is so abundant. In the same century, and in fact in the same generation, we have the Desargues-Pascal discovery of projective geometry and the Pascal-Fermat discovery of the principles of a mathematical theory of chance. But these phenomena were by no means confined to the seventeenth century. The eighteenth century had the Newton-Leibnitz incident; the nineteenth century witnessed the almost simultaneous discovery by Wessel, Argand, and Gauss of an interpretation of the complex magnitudes; the nearly simultaneous conception of non-Euclidean geometry by Lobatchevski, Bolyai, and Gauss; and late in the century the Cantor-Dedekind formulation of the continuum.

Similar examples exist in other sciences. The same conception arises in the brains of two or even more men at practically the same time. In many cases the men are separated by thousands of miles, belong to entirely different nationalities, and are not even aware of each other's existence; and the differences in temperament, environment, and outlook of two such men as Descartes and Fermat are striking. How can we account for this strange phenomenon? It seems as if the accumulated experience of the race at times reaches a stage where an outlet is imperative, and it is merely a matter of chance whether it will fall to the lot of a single man, two men, or a whole throng of men to gather the rich overflow.

Neither Fermat nor Descartes realized the full significance of their discovery. Both were interested in the creation of a unifying principle in geometry: Fermat from the standpoint of a pure mathematician, Descartes from that of a philosopher. Greek geometry, which found its final expression in the works of Euclid and Apollonius, did not possess such unity: every theorem, every construction seemed more like an artistic creation than the application of general principles. What ideas lay concealed behind this or that construction? Why were certain problems constructible by straight-edge only, while others required the compass as well, and still others would not yield even to the ingenuity of the Greeks, past masters of the ruler and compass? These and similar questions agitated the mathematical minds of that period, Fermat and Descartes among others.

They sought the clue in algebra; so they proceeded to *algebrise* geometry, and analytic geometry was the result. They laid the foundation of the procedure by means of which a problem in geometry may be reduced to the prosaic manipulations of algebra. Thus, the famous problems of antiquity, which began in legendary splendor and were throughout the ages a source of fascination to many mathematicians of rank, were now disposed of by Descartes in the matter-of-fact statement that any problem which leads to an equation of the first degree is capable of a geometrical solution by straight-edge only; that a straight-edge-compass construction is equivalent to the solution of a quadratic equation; but that if a problem leads to an *irreducible* equation of a degree higher than the second, its geometrical solution is not possible by means of a ruler and compass only.

Descartes (and the same of course applies to Fermat) was not aware that he was laying the foundations of a new mathematics; his avowed purpose was to systematize the geometry of the ancients. This, indeed, was the rôle which the seventeenth century played in the history of mathematics: it was *the age of liquidation* of the antique mathematical culture. I see in the work of Galileo, Fermat, Pascal, Descartes, and others the consummation of an historical process which could not reach its climax in a period of general decline. Roman indifference and the long Dark Ages of religious obscurantism prevented a resumption of this process for fifteen hundred years.

At the same time, by clearing away the débris of ancient mathematics, the genius of these men prepared the ground for the new. The essential characteristics of modern mathematical thought are *the permanence of formal laws* and the *principle of correspondence.* The first led to the generalized number concept, the second permitted the establishment of the kinship between seemingly remote and dissimilar concepts. Though Descartes lacked even an implicit understanding of these two fundamental principles of modern mathematics, his analytic geometry contained all that was necessary for the development of these principles.

Here was an algebra which implicitly admitted irrationals on terms of equality with rational magnitudes. It was being applied to the classical problems of geometry: by direct and methodical processes this algebra was yielding the same results that the Greeks—committed as they were to the utmost rigor, and hampered as they were by the fear of irrational numbers and infinity—obtained by ingenious but unmethodical schemes.

This fact itself gave to the deductions of Descartes a tremendous pragmatic strength, for nothing succeeds like success.

In the second place analytic geometry was the first historical example of a kinship established between two branches of mathematics not only remote in nature, but known, from the very beginning of mathematics, to be in direct conflict: *arithmetic* and *geometry*. This last phase was not apparent to Fermat, Descartes or their contemporaries, but in the course of the next two hundred years it was destined to exert the greatest influence on the development of mathematical thought.

I said in the preceding chapter that Descartes implicitly assumed that a complete correspondence existed between the real numbers and the points of a fixed axis. He assumed more than that: tacitly, because it seemed so natural as to go without saying, he accepted it as axiomatic that *between the points of a plane and the aggregate of all the pairs of real numbers there can be established a perfect correspondence.* Thus the Dedekind-Cantor axiom, extended to two dimensions, was tacitly incorporated in a discipline which was created two hundred years before Dedekind or Cantor saw the day. This discipline became the proving-grounds for all the achievements of the following two centuries: the calculus, the theory of functions, mechanics, and physics. Nowhere did this discipline, analytic geometry, strike any contradiction; and such was its power to suggest new problems and forecast the results that wherever applied it would soon become an indispensable tool of investigation.

Take two perpendicular axes, assign a sense on each; then any point in the plane of the axes can be represented by two numbers. Each one of these may be positive, zero, or negative, rational or irrational. These numbers are the *measures* of the distances of the given point to the axes of reference, preceded by plus or minus, depending on which of the four *quadrants,* determined by the axes, the point may be.

The principle is so simple, so natural, that it is difficult to believe that it took three thousand years to discover it. The phenomenon is as striking as the one presented by the principle of position in numeration. The latter is implicitly contained in the structure of our number language, and yet was not discovered for five thousand years. The former is a direct consequence of the symmetrical structure of our body and has been used in describing the mutual position of bodies since time immemorial. It would seem indeed that it is only necessary to attach a quantitative meaning to the ideas of *right* and *left,* of *back* and *forth,* of *up* and *down,* to have a full-fledged *coordinate* geometry.

And we find this principle utilized from the earliest days; we find it in the ancient fairy tales which describe the location of a treasure by instructing the seeker to take so many steps to the East, and then so many to the North; we find that the Egyptian surveyors explicitly applied it by tracing a South-North and an East-West line and referring any object to these axes.

The transition from this practical procedure to analytic geometry depended of course on the creation of zero and of the negative number concept. But these were known in Europe since the days of Fibonacci. Why is it, then, that the *coordinate principle* did not occur to mathematicians earlier? The answer may be found in the tremendous influence which Greek opinion exercised upon European thought. The emancipation of number from the inhibitions imposed on it by the Greeks was not as easy a task as it may appear to us today.

Cartesian geometry assigns to every point in the plane two real numbers, and to any pair of real numbers a point in the plane. It identifies the aggregate of real couples with the points in the plane. From this it is but one step to regarding the point as a *number-individual,* as a single number. Yet this step, too, lagged for nearly two centuries.

In 1797 an obscure Norwegian surveyor by the name of Wessel presented before the Danish Academy of Sciences a report on the geometrical interpretation of complex quantities. This report passed unnoticed, and only one hundred years later did it become known to the scientific world. In the same year, 1797, the twenty-year-old Gauss was defending his doctor's thesis on the fundamental theorem of algebra, in which he implicitly used a geometrical representation of the complex domain. In 1806 Robert Argand, an obscure Parisian bookkeeper, Swiss by birth, published an essay on the geometrical interpretation of the complex. This again passed unnoticed until about ten years later when it was republished in a prominent mathematical journal. Finally, in 1831 Gauss, in the essay quoted before, formulated with precision *the mathematical equivalence of plane Cartesian geometry with the domain of the complex number.*

According to this formulation, which is essentially that of Wessel and Argand, a real number represents a point on the x-axis of the Cartesian diagram. If a be such a real number (see accompanying figure) and A its representative point on the x-axis, then multiplication by i is equivalent to revolving the *vector OA* through a right angle counter-clockwise; the number ia is therefore represented by a point A' on the y-axis. If this again be multiplied by i we obtain $i^2a = -a$, which represents the point A'' on the x-axis, etc., etc. Four successive rotations

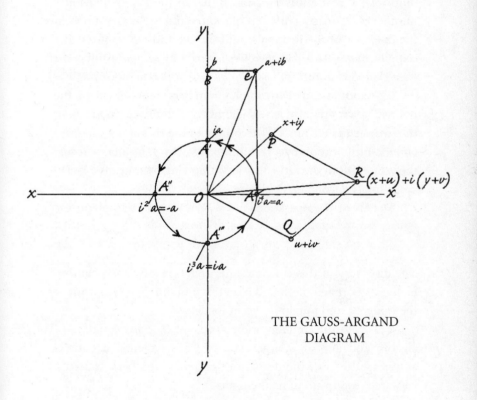

THE GAUSS-ARGAND
DIAGRAM

through a right angle each, will bring the point back to its original position. This is then the geometrical interpretation of the relations which were mentioned on page 192.

Furthermore the addition $a + ib$ is interpreted as the *composition* of the vectors *OA* and *OB*, where *A* is the representative point of the real number *a*, and *B* that of the *purely imaginary number ib*. Consequently $a + ib$ represents the extremity *C* of the diagonal of the rectangle built on *OA* and *OB* as sides. *The complex number* a + ib *is therefore identified with that point of the Cartesian diagram which has* a *for abscissa and* b *for ordinate.*

The addition of two complex numbers, represented by the points *P* and *Q* respectively, is identified with the composition of the two vectors *OP* and *OQ* according to the *rule of the parallelogram*. Multiplication by a real number, say by 3, means *stretching* the vector *OP* in the ratio 3 to 1. Multiplying by *i* means a right-angle rotation counter-clockwise, etc., etc.

Details of the operations are shown on the example in the figure.

The discovery of this concrete interpretation gave the phantom beings of Bombelli flesh and blood. It took the *imaginary* out of the *complex*, and put an *image* in its place.

This, and the contemporaneous proof that any equation of algebra and any of a large class of transcendental equations admit solutions in the domain of complex numbers, caused a veritable revolution in mathematics.

In the field of analysis Cauchy, Weierstrass, Riemann, and others extended the whole apparatus of infinite processes to the complex domain. The theory of functions of a complex variable was thus established, with all its far-reaching consequences for analysis, geometry, and mathematical physics.

In the field of geometry, Poncelet, Von Staudt, and others took the complex number as a point of departure to establish a general projective geometry; Lobachevski, Bolyai, Lie, Riemann, Cayley, Klein, and many others opened up the rich field of non-Euclidean geometries. The applications of the complex number to infinitesimal geometry eventually grew into the absolute differential geometry which is the foundation of the modern Relativity theory.

In the theory of numbers Kummer invented the method of complex divisors, which he called *ideal* numbers, and thereby advanced the Fermat problem and related questions to an undreamed-of stage.

This tremendous success encouraged generalizations: these proceeded in two directions. In the first place, the question was considered, whether complex units could be used which follow other laws than $i^2 = -1$. A great deal of work has been done in this direction, but it is of no special relevance to this survey. In the second place, it was natural to inquire whether the points in space of three dimensions could also be considered as number-individuals: out of this inquiry grew a new discipline which eventually became the *Vector Analysis* of today, that plays such a fundamental rôle in modern mechanics. Another outgrowth of this inquiry was the theory of *quaternions* founded by Hamilton, and the related Grassmann theory of *extensive magnitudes*.

These generalizations disclosed the important fact that extensions beyond the complex number domain are *possible only at the expense of the principle of permanence*. The complex number domain is the last frontier of this principle. Beyond this either the commutativity of the operations or the rôle which zero plays in arithmetic must be sacrificed.

This forced the investigation of the properties of operations generally. The principle of permanence was then extended by dropping some of the restrictions. The result was the construction of a far-reaching theory of matrices, a theory in which a whole array of elements is regarded as a number-individual. These "filing cabinets" are added and multiplied, and a whole calculus of matrices has been established which may be regarded as a continuation of the algebra of complex numbers. These abstract beings have lately found a remarkable interpretation in the quantum theory of the atom, and in man's other scientific fields.

Such in rough outline is the story of the complex magnitude. For centuries it figured as a sort of mystic bond between reason and imagination. In the words of Leibnitz,

> "The Divine Spirit found a sublime outlet in that wonder of analysis, that portent of the ideal world, that amphibian between being and not-being, which we call the imaginary root of negative unity."

Others regarded the complex as an empty play upon symbols which for some unaccountable reason led to actual results. They were useful, and that justified their existence, as the means justifies the end. They supplied the method and anticipated the result of many otherwise impregnable problems. So these phantoms were often invoked, but never without misgiving.

And then came the day when it was shown that these phantom creatures of Bombelli were not phantoms at all, but that they had just as concrete an existence as any real number. Furthermore, these complex beings led a sort of double existence: on the one hand they obeyed all the laws of arithmetic and were for this reason *bona fide* numbers; on the other hand they found a complete incarnation in the points of the plane. They were therefore ideally adapted as an instrument for translating into the language of number the intricate geometrical interrelations between the configurations of the plane.

When this was realized, the *arithmetization* of *geometry,* which was unintentionally begun by Fermat and Descartes, became an accomplished fact. And so it was that the complex number, which had its origin in a *symbol for a fiction,* ended by becoming an indispensable tool for the formulation of mathematical ideas, a powerful instrument for the solution of intricate problems, a means for tracing kinships between remote mathematical disciplines.

Moral: FICTION IS A FORM IN SEARCH OF AN INTERPRETATION.

CHAPTER 11

The Anatomy of the Infinite

"The essence of mathematics is its freedom."
—Georg Cantor

To measure the *plurality* of an infinite collection seems a bizarre idea at first. Yet even those least conversant with mathematical ideas have a vague feeling that there are infinities *and* infinities—that the term infinity as attached to the natural sequence of numbers and the term infinity used in connection with points on a line are essentially different.

This vague idea which we have the "contents" of an infinite collection may be compared to a net. It is as if we threw a net of unit mesh and so singled out the whole numbers, all other numbers passing through. We then took a second net of mesh 1/10; then a third of mesh 1/100; and continuing this way we gathered up more and more of the rational numbers. We conceive of no limit to the refinement of this process, for no matter how fine a net we may cast, there is yet a finer that *could be* cast. Give free rein to your imagination and you can picture this ultra-ultimate net so compact, of mesh so fine, as to be able to gather up *all* the rational numbers.

It is when we push this analogy to the extreme and begin to regard this limiting net as something fixed, frozen as it were, that we strike all the difficulties so skilfully brought out by Zeno. Here, however, we are concerned with another difficulty.

215

This ultra-ultimate rational net, even if it could be materialized, would still be incapable of gathering *all numbers*. A net still more "compact" is necessary to take care of the irrationals of algebra; and even this "algebraic" net would be incapable of gathering the transcendental numbers. And thus our intuitive idea is that the rational number domain is more compact than the natural; that the algebraic numbers are arranged in still denser formation; and that finally the real number domain, the arithmetic continuum, is the *ultra-dense medium,* a medium without gaps, a network of mesh zero.

If then we are told for the first time that Georg Cantor made an actual attempt to classify infinite collections and to endow each with a number representative of its plurality, we naturally anticipate that he succeeded in finding a measure of this variable compactness.

And just because such were our anticipations, the achievement of Cantor has many surprises in store for us, some of these so striking as to border on the absurd.

The attempt to measure the compactness of a collection by means of nets is doomed to failure because it is physical in principle, and not arithmetical. It is not arithmetical, because it is not built on the principle of correspondence on which all arithmetic rests. The classification of the *actually infinite,* i.e., of the various types of plurality of infinite collections, if such a classification is at all possible, must proceed on the lines along which pluralities of finite collections were classified.

Now we saw in the opening chapter that the notion of *absolute plurality* is not an inherent faculty of the human mind. The genesis of the natural number, or rather of the cardinal numbers, can be traced to our matching faculty, which permits us to establish correspondence between collections. The notion of *equal-greater-less* precedes the number concept. We learn *to compare* before we learn *to evaluate*. Arithmetic does not begin with numbers; it begins with *criteria*. Having learned to apply these criteria of equal-greater-less, man's next step was to devise models for each *type* of plurality. These models are deposited in his memory very much as the standard meter is deposited at the Bureau of Longitudes in Paris. One, two, three, four, five …; we could just as well have had: I, wings, clover, legs, hand … and, for all we know, the latter preceded our present form.

The principle of correspondence generates the integer and through the integer dominates all arithmetic. And, in the same way, before we can measure the plurality of infinite collections, we must learn to compare them. How? In the same way that this was accomplished for finite collections. The matching process, which performed such signal services in finite arithmetic, should be extended to the *arithmetic of the infinite:* for the elements of two infinite collections might also be matched one by one.

The possibility of establishing a correspondence between two infinite collections is brought out in one of the dialogues of Galileo, the first historical document on the subject of infinite aggregates. I reproduce this dialogue verbatim from a book entitled "Dialogs Concerning the New Sciences," which appeared in 1636. Three persons participate in the dialogues. Of these Sagredo represents the practical mind, Simplicio one who is trained in scholastic methods, while Salviati is obviously Galileo in person.

> *Salviati*: This is one of the difficulties which arise when we attempt with our finite minds to discuss the infinite, assigning to it those properties which we give to the finite and limited; but

this I think is wrong for we cannot speak of infinite quantities as being the one greater or less than or equal to the other. To prove this, I have in mind an argument which for the sake of clearness, I shall put in the form of questions to Simplicio, who raised this difficulty.

I take it for granted that you know which of the numbers are squares and which are not.

Simplicio: I am quite aware that a square number is one which results from the multiplication of another number by itself; thus, 4, 9, etc., are square numbers which come from multiplying 2, 3, etc., by themselves.

Salviati: Very well; and you also know that just as the products are called squares, so the factors are called sides or roots; while on the other hand those numbers which do not consist of two equal factors are not squares. Therefore, if I assert that all numbers including both squares and non-squares are more than the squares above, I shall speak the truth, shall I not?

Simplicio: Most certainly.

Salviati: If I ask how many squares there are, one might reply truly that there are as many as the corresponding numbers of roots, since every square has its own root and every root its own square, while no square has more than one root and no root more than one square.

Simplicio: Precisely so.

Salviati: But if I inquire how many roots there are, it cannot be denied that there are as many as there are numbers, because every number is a root of some square. This being granted, we must say that there are as many squares as there are numbers, because they are just as numerous as their roots, and all the numbers are roots. Yet, at the outset, we have said that there are many more numbers than squares, since the larger portion of them are not squares. Not only so, but the proportionate number of squares diminishes as we pass to larger numbers. Thus up to 100, we have ten, i.e., the squares constitute one-tenth of all numbers; up to 10,000 we find only one hundredth part to be squares, and up to a million only one thousandth part; and, yet, on the other hand, in an infinite number, if one could conceive of such a thing, he would be forced to admit that there are as many squares as there are numbers all taken together.

Sagredo: What, then, must one conclude under such circumstances?

Salviati: So far as I see, we can only infer that the number of squares is infinite and the number of their roots is infinite; neither is the number of squares less than the totality of all numbers, nor the latter greater than the former; and finally the attributes "equal," "greater," and "less" are not applicable to infinite, but only to finite quantities.

When, therefore, Simplicio introduces several lines of different lengths and asks how is it possible that the longer ones do not contain more points than the shorter, I answer him that one line does not contain more, or less, or just as many, points as another, but that each line contains an infinite number.

The paradox of Galileo evidently left no impression on his contemporaries. For two hundred years nothing was contributed to the problem. Then in 1820 there appeared a small tract in German by one Bolzano, entitled "The Paradoxes of the Infinite." This, too, attracted little attention; so little indeed, that when fifty years later the theory of aggregates became the topic of the day, few mathematicians knew who the man was.

Today Bolzano's contributions are of a purely historical interest. While it is true that he was the first to broach the question of the *actually infinite* he did not go far enough. Yet, due honor must be given the man for creating the all-important concept of the *power* of an aggregate of which I shall speak shortly.

The modern *theory of aggregates* begins with Georg Cantor. His essay, which laid the foundation of this new branch of mathematics, appeared in 1883 under the title "On Linear Aggregates." This essay was the first to deal with the actually infinite as with a definite mathematical being. The following passage from this essay will bring out clearly Cantor's approach to the problem:

> "It is traditional to regard the infinite as the indefinitely growing or in the closely related form of a convergent sequence, which it acquired during the seventeenth century. As against this I conceive the infinite in the definite form of something consummated, something capable not only of mathematical formulations, but of definition by number. This conception of the infinite is opposed to traditions which have grown dear to me, and it is much against my own will that I have been forced to accept this view. But many years of scientific speculation and trial point to these conclusions as to a logical necessity, and for this reason I am confident that no valid objections could be raised which I would not be in position to meet."

To appreciate the great courage which it required to break so openly with the traditions of the past, we must understand the universal attitude of Cantor's generation towards the *actually infinite*. For this purpose, I quote from a letter of the great Gauss to Schumacher, which, although written in 1831, set the tone of the mathematical world for the next half-century:

> "As to your proof, I must protest most vehemently against your use of the infinite as something consummated, as this is never permitted in mathematics. The infinite is but a figure of speech; an abridged form for the statement that limits exist which certain rations may approach as closely as we desire, while other magnitudes may be permitted to grow beyond all bounds ...

"... No contradictions will arise as long as Finite Man does not mistake the infinite for something fixed, as long as he is not led by an acquired habit of the mind to regard the infinite as something bounded."

Gauss's ideas on the subject were universally shared, and we can imagine what a storm Cantor's open defiance raised in the camp of the orthodox. Not that the actually infinite was not in one guise or another used in the days of Cantor, but that in such matters the traditional attitude was like that of the Southern gentleman with respect to adultery: he would rather commit the act than utter the word in the presence of a lady.

It was fortunate for Cantor that mature reflection had thoroughly steeled him to face the onslaught, because for many years to come he had to bear the struggle alone. And what a struggle! The history of mathematics has not recorded anything equal to it in fury. The story beginnings of the theory of aggregates show that even in such an abstract field as mathematics, human emotions cannot be altogether eliminated.

Cantor begins where Galileo left off. Yes, it is possible to establish a correspondence between two infinite collections, even if one is but a part of the other! For precision, therefore, let us say that two collections, *finite* or *infinite,* are equivalent, or have the same *power,* if they can be matched element for element. If two collections are of *different power,* then the matching process will exhaust one, but there will still remain unmatched elements in the other. In other words, the first may be matched with *a part* of the second, but the second cannot be matched with any part of the first. Under the circumstances, we say that the second aggregate is of a power *greater* than the first.

If (A) and (B) are two *finite* collections, each containing the same number of elements, then obviously they have the same power; and, conversely, if (A) and (B) are *finite* collections of equal power, they have also the same *cardinal* number. If (A) and (B) are of unequal power, then to the greater power corresponds also the greater cardinal number. *For finite collections, therefore, the concept of power can be identified with that of cardinal number.* Now, since in the *arithmetic of the finite* the term power can be identified with cardinal number, it is natural to inquire whether it is possible to identify the powers of infinite collections with numbers of a higher order, *transfinite* numbers as it were, and by means of this new concept create a *transfinite arithmetic,* an arithmetic of the infinite.

If we proceed along the lines suggested by the beginnings of finite arithmetic, we must seek model-aggregates, each model representative of some typical plurality. Such models are close at hand: the natural sequence, the rational domain, the field of algebraic numbers, the arithmetic continuum—all these infinite collections which have grown so familiar to us through constant use are admirably adapted as standards of comparison. Let us then endow these standard collections with symbols, and have these symbols play the same rôle in a transfinite arithmetic as their counterparts, the finite cardinal numbers, 1, 2, 3 ..., play in the arithmetic of the finite.

These symbols Cantor calls the *transfinite cardinals*. He orders them in a "sequence" of growing power; he defines the operations of addition, multiplication, and potentiation upon these abstract beings; he shows how they combine among themselves and with finite cardinals. In fine, these illusory creatures of Cantor's genius possess so many of the properties of finite magnitudes that it seems altogether proper to confer upon them the title "number." But one all-important property they do not possess, and that is *finitude.* This last statement sounds like a truism, and yet it is not intended in any spirit of triviality. All the paradoxical propositions which I am about to present derive from the fact that these mathematical beings, which have all the appearances of numbers, are deprived of some of the most rudimentary attributes of common number. One of the most striking consequences of this definition is that a part of a collection is not necessarily less than the whole: it may be equal to it.

The part may have the power of the whole. This sounds more like theology than mathematics. And, indeed, we find this idea being toyed with by many a theologian and near-theologian. In the Sanskrit manuals, where religion is so delightfully intermixed with philosophy and mathematics and sex instruction, such ideas are quite usual. Thus Bhaskarah in speculating on the nature of the number 1/0 states that it is "like the Infinite, Invariable God who suffers no change when old worlds are destroyed or new ones created, when innumerable species of creatures are born or as many perish."

"The part has the power of the whole." Such is the essence of Galileo's paradox. But while Galileo dodged the issue by declaring that "the attributes of equal, greater, and less are not applicable to infinite, but only to finite quantities," Cantor takes the issue as a point of departure for his theory of aggregates.

And Dedekind goes even further: to him it is characteristic of all infinite collections that they possess parts which may be matched with the whole. For purposes of illustration, consider any infinite sequence ordered and labeled accordingly. Now drop any finite number of terms in the beginning and re-label the curtailed sequence. For every term of the second, there was a term in the original sequence of the same rank, and vice versa. The correspondence is, therefore, complete and the two sequences possess the same power; and yet it cannot be denied that the second is but a part of the first. Such a phenomenon is possible only in infinite collections, for it is characteristic only of *finite* collections that the whole is never equal to a part.

But let us return to the Cantor theory. The symbol a will designate the power of the aggregate of natural numbers. Any aggregate which possesses the power a will be called *denumerable*. The sequence of perfect squares, used in Galileo's argument, is such a denumerable aggregate. But so is every other sequence, for the mere fact that we can assign a rank to any term shows that there is a perfect correspondence between the sequence and the natural numbers. The even numbers, the odd numbers, any arithmetic progression, any geometric progression, any sequence at all, is denumerable.

What is more, if any such sequence is imagined removed from the domain of natural numbers, the remaining aggregate is still infinite and still denumerable; and this is why there is no hope of reducing the power of a denumerable set by a *thinning-out* process. We may, for instance, remove all even numbers, then all remaining multiples of 3, then all remaining multiples of 5. We may continue this process indefinitely without affecting the power of what remains.

In the language of Cantor, there is *no smaller transfinite number* than the number a, which measures the plurality of any denumerable infinite collection.

But if there be no hope of obtaining a smaller transfinite by thinning out the natural sequence, could we not increase the power by a process of filling-in? It would appear, indeed, that the power of the rational domain, which is everywhere dense, should be greater than that of the natural sequence which is discrete. Here again our intuition leads us array, for Cantor proves to us that the rational aggregate is also denumerable. To prove this, it is only necessary to show that the rational numbers can themselves be arranged in a sequence, by assigning to each rational number a definite rank. This is what Cantor actually does. We can get a general idea of the method by considering it geometrically.

In the accompanying figure we have two sets of parallel lines, at right angles to each other. We identify any line of the horizontal set by a whole number y, the number y taking on all integral values from $-\infty$ to $+\infty$; similarly for the number x, which identifies the vertical lines. Now we label any joint of the infinite lattice we have thus erected by the two numbers which identify the vertical and the horizontal lines there intersecting. Thus the symbol (y, x) identifies a determinate joint in our lattice work, and conversely any point is capable of such a representation.

We shall show that the totality of these joints form a denumerable aggregate. To prove this striking fact, it is sufficient to draw the spiral-like polygonal line as in the figure and follow the joints in the order in which they appear on the diagram.

On the other hand, we can identify the symbol (y, x) with the fraction y/x. But if we do this, it is obvious that we cannot label *all* our joints with distinct rational numbers. In fact, all joints situated on the same line through the origin represent one and the same rational number, as it is easy to convince oneself. To eliminate this ambiguity, *we agree to count each fraction only the first time it occurs.* These points form the sequence:

$$1, 0, -1; -2, 2, +\tfrac{1}{2}, -\tfrac{1}{2}; -\tfrac{3}{2}, -3, +3, +\tfrac{3}{2}, +\tfrac{2}{3}, +\tfrac{1}{3}, \dots$$

(See figure, page 226.)

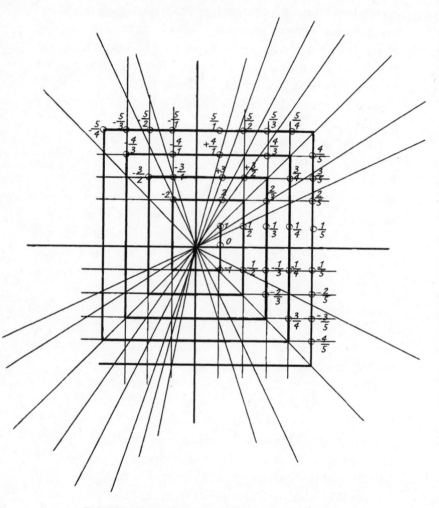

DOCUMENTING THE RATIONAL DOMAIN

Here all rationals are represented, and each rational occurs in the sequence but once. *The domain of rational numbers is therefore denumerable.*

But, the reader may exclaim, this stands in direct contradiction to our notion of compactness, according to which no rational number has a successor. Between any two rational numbers we can insert an infinity of others; but here we have actually established a succession! The answer to this is that, while we have here a true succession, it is not of the same type as the natural succession 1, 2, 3, ... which is arranged according to magnitude. We succeeded in enumerating the rational numbers because in the new arrangement we were not obliged to preserve the order of magnitude. We have obtained *succession at the expense of continuity.*

We see that it is essential to discriminate between two kinds of equivalence. From the standpoint of *correspondence,* two collections are equivalent if they can be matched element by element. From the standpoint of *order,* this is also indispensable. But for complete equivalence, for *similarity,* it is necessary in addition that the matching process should not destroy the order of arrangement: i.e., if in the collection (A) the element a preceded the element a', then in the collection (B) the corresponding element b must precede b'. The aggregate of rational numbers arranged according to magnitude, and the spirally arranged aggregate by means of which we denumerated the rational numbers, are equivalent from the standpoint of correspondence, but not from the standpoint of order. In other words, they have the same cardinal number, a, but are of different *ordinal types.*

Hence Cantor proposed a theory of *ordinal types* which form the counterpart of the ordinal numbers of finite arithmetic. Theorem, however, we had the fundamental property that any two collections with equal cardinal numbers had also the same ordinal number, and to this we owed the facility with which we passed from one to the other. But in the Cantor arithmetic of the infinite, two aggregates may be measured by the same cardinal number and yet be *ordinally distinct,* or, as Cantor says, *dissimilar.*

Thus mere compactness is no obstacle to denumeration, and the filling-in process does not affect the power of an aggregate any more than the thinning-out process did. So the next Cantor deduction is somewhat less of a shock to us; this states that *the aggregate of algebraic numbers is also denumerable.* Cantor's proof of this theorem is a triumph of human ingenuity.

He commences by defining what he calls the *height* of an equation. This is the sum of the *absolute values* of the coefficients of the equation, to which is added its degree diminished by 1. For instance, the equation $2x^3 - 3x^2 + 4x - 5 = 0$ has a height $h = 16$, because $2 + 3 + 4 + 5 + (3 - 1) = 16$.

He proves next that there are but a *finite* number of equations which admit any positive integer h for height. This permits us to order *all* algebraic equations in groups of increasing height; it can be shown that there is only one equation of height 1; three of height 2; twenty-two of height 3, etc.

Now, within each group of given height, we can order the equations by any number of schemes. For instance, we can combine all equations of the same degree into one sub-group and arrange each sub-group according to the magnitude of the first coefficients, for those which have the same first coefficient according to the second, etc., etc.

Any such scheme would allow us to order all algebraic equations into a hierarchy and thus enumerate them, i.e., assign to each equation a rank. Now to every one of these equations may correspond one or more real roots, but the number is always finite and in fact cannot exceed the degree of the equation—and therefore cannot exceed the height; these roots can again be arranged according to their magnitude. Now if we consider the scheme as a whole, we shall surely find repetitions, but as in the case of the rational numbers, these repetitions can be avoided by agreeing to count any algebraic number only the first time it occurs in the process.

In this manner we have succeeded in assigning to any algebraic number a rank in the hierarchy, or in other words, we have *denumerated the aggregate of algebraic numbers.*

By this time the suspicion will have grown upon the reader that perhaps *all aggregates are denumerable.* If such were the case there would be but one transfinite, and what was true for the rational and the algebraic aggregates would be generally true even of the continuum. By some artifice, such as Cantor's height, any infinite collection could be arranged into a hierarchy and thus enumerated. Such, indeed, was Cantor's idea in the early stages of his work: *to enumerate the real numbers* was one of the points of his ambitious program; and the theory of transfinite numbers owes its birth to this attempt to "count the continuum."

That such is not the case that it is not possible to arrange all real numbers in a denumerable sequence, was known to Cantor as early as 1874. However, the proof of it did not appear until 1883. I cannot go into the details of this demonstration, the general principle of which consists in assuming that all real numbers have been erected into a hierarchy, and then in showing, by a method which we now call the *diagonal procedure,* that it is possible to exhibit other numbers which while real are not among those which have been enumerated.

One sidelight on this proof has an important historical bearing. The reader will remember Liouville's discovery of transcendentals. This existence theorem of Liouville was re-established by Cantor as a sort of by-product of his theorem that the continuum cannot be denumerated. The relative wealth of the two domains, the algebraic and the transcendental, a question which to Liouville had only a vague meaning, was now formulated by Cantor in full rigor. He showed that whereas the algebraic domain has the power a of the aggregate of natural numbers, the transcendentals possess the *power c of the continuum.* Thus the contention that there are incomparably *more* transcendentals than algebraic numbers acquires a true significance.

And here, too, in this domain of real numbers the part may have the power of the whole; in the quaint language of Galileo "the longer line contains no more points than the shorter." In fact, a segment of a line, no matter how short, has the same power as the line indefinitely extended, an area no matter how small has the power of the infinite space of three dimensions. In short: *parceling or piecing together can no more affect the power of an aggregate than thinning out or filling in.*

At this point our intuition again whispers a suggestion. How about these manifolds of higher dimensions: the complex number domain, which we identified with the set of points in the plane; the points in space; the vectors and quaternions; the tensors and the matrices, and other intricate complexes which mathematicians manipulate as though they were individuals, subject to the laws of operations on numbers, but which cannot be represented in a continuous manner as points on a line? Surely these manifolds should have a power higher than that of the *linear* continuum! Surely there are more points in space of three dimensions, this universe extending indefinitely in all directions, than on a segment of a line one inch long!

This, too, may have been an early idea of Cantor. But he proved conclusively that here too our intuition leads us astray. The infinite manifold of two or three dimensions, the mathematical beings which depend on a number of variables greater even than three, any number in fact, still have no greater power than the *linear continuum*. Nay, even could we conceive of a variable being whose state at any instant depended on an infinite number of independent variables, a being which "lived" in a *world of a denumerable infinite of dimensions,* the totality of such beings would still have a power not greater than that of the linear continuum not greater than a segment one inch long.

This statement strikes us as being in such direct contradiction with our ideas of dimension as to be absurd. Such was, indeed, the opinion of many when Cantor first announced it, and there were first-class minds who took it warily, to say the least. But Cantor's proof of this fundamental proposition is so simple that even a bright child can see it.

I shall illustrate the statement as it applies to the points in the plane: the reader will see that the argument is perfectly general. Since the points within a segment of length 1 have the same power as the indefinite line, and the points within the square of side 1 the same power as the indefinite plane, it will be sufficient to show that a one-to-one correspondence can be established between this square and this segment.

Now any point P within this square OAFB of the accompanying figure can, as we saw, be represented by means of two coordinates x, y. These latter are real numbers not exceeding 1 and can be exhibited as proper decimal fractions. These fractions can be always regarded as interminable for even if terminable they may be rendered interminable by an adjunction of zeros behind

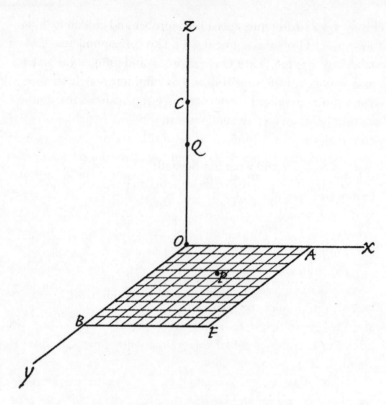

MAPPING A SQUARE ON A LINE

the last significant figure. Let us then write these decimal fractions in the form:

$$x = . a_1 \mid a_2 \mid a_3 \mid a_4 \mid a_5 \mid a_6 \mid \ldots$$
$$y = . b_1 \mid b_2 \mid b_3 \mid b_4 \mid b_5 \mid b_6 \mid \ldots$$

Now let us form a third decimal fraction z by taking alternately the figures x and y

$$z = . a_1 \mid b_1 \mid a_2 \mid b_2 \mid a_3 \mid b_3 \mid a_4 \mid b_4 \mid \ldots$$

This fraction again represents a real number and we can exhibit it as a point Q on the segment OC. The correspondence thus established between P and Q is reciprocal and unique; for given x and y we can always form z, and in only one way; and conversely, the knowledge of z permits us to reconstruct the numbers x and y, and therefore the point P.

What lies between and what lies beyond?

There is nothing in the Cantor theory to preclude the possibility of a transfinite number greater than a, but less than c. Yet all known point-aggregates are either denumerable, like the rational or algebraic number domains, or else, like the transcendentals, have the power of the arithmetic continuum. All attempts to erect an artificial point-aggregate which would be "mightier" than the natural sequence, but less "mighty" than the aggregate of the points on a line, have so far not been crowned with success.

On the other hand, aggregates are known which have a power greater than c. Among these there is the so-called *functional* manifold, i.e., the *totality of all correspondences* which can be established between two continua. This totality cannot be matched with the natural numbers. The corresponding cardinal number is denoted by f. Again, there is nothing in the theory that would preclude the existence of cardinals between c and f, and yet no aggregate has yet been discovered of power less than f, but greater than c.

And beyond f there are still greater cardinal numbers. The same diagonal procedure which permits us to derive the functional "space" from the continuum can be used to derive from the functional space a *superfunctional* which *cannot* be matched with the aggregate of correspondences. Aggregates of higher and higher power can be thus erected, and the process *cannot conceivably be terminated.*

So pushed to its ultimate border, the Cantor theory asserts that there is *no last transfinite number.* This assertion is strangely similar to the other: there is *no last finite number.* Yet the latter was admittedly an assumption, the fundamental assumption of *finite* arithmetic, whereas the analogous statement in the arithmetic of the *infinite* seems to be the logical conclusion of the whole theory.

There is no last transfinite! The proposition sounds innocent enough, and yet it contained within itself an explosive which nearly wrecked the whole theory, and that at a time when Cantor, having overcome the powerful resistance of his first opponents, had all reason to believe that his principles had emerged triumphant. For almost simultaneously a series of "phenomena" were uncovered which, while seemingly different in character, indicated that something was wrong. The Italian Burali-Forti, the Englishman Bertrand Russell, the German König, and the Frenchman Richard unearthed antinomies and paradoxes, which bear their respective authors' names. Again, the question was raised as to the validity of the Cantor methods and deductions, as to the legitimacy of the use of the actually infinite in mathematics.

It would take me too far to go into detail about the nature of the contradictions discovered. Heterogeneous though these paradoxes are, they all seem to hinge on the questions how the word *all* should be used in mathematics, if it is to be used at all. If this word can be used freely in connection with *any* conceivable acts of the mind, then we can speak of the *aggregate of all aggregates.* If now this is an aggregate in the Cantor sense, then it too must possess a cardinal number. This transfinite number is the "greatest conceivable," for can we conceive of an aggregate mightier than the aggregate of all aggregates? This cardinal is, therefore, the *last transfinite,* the truly ultra-ultimate step in the evolution of the abstraction which we call number! And yet there is no last transfinite!

Much water has passed under the bridge since these paradoxical questions were first raised; many solutions have been offered, thousands of essays have been written on both sides of the question, much sarcasm indulged in by the Cantorians and their opponents. Yet the question remains wide open. Cantor found mathematics undivided; he left it split into two contending camps.

To present the "platforms" of these opposing mathematical "parties" with the simple means at my disposal, and pledged as I am to avoid technicalities, is impossible. And yet I should fall short of my own program were I to pass up entirely this most vital issue of modern mathematics. So I shall simply and briefly state the dilemma as voiced by the chief representatives of the opposing camps.

On the side of the "formalists" are Hilbert, Russell, Zermelo. While defending Cantor they are "mensheviki" in the sense that they are attempting to save his *minimum* program. They admit that the unrestricted use of the words "all," "aggregate," "correspondence," and "number," is inadmissible. But the solution lies not in the complete negation of the theory of aggregates but in the remolding of the theory along the lines of pure reason. We must devise a body of axioms which could serve as the basis of the theory and, to make sure that we are not again led astray by our institution, we must construct a purely *formal*, logically consistent schematic outline of such a body, a mere skeleton without content. Having erected such a comprehensive, consistent system, we shall base the arithmetic of the infinite upon it as a foundation, secure in our conviction that no paradox or antimony will ever arise again to disturb our peace of mind. Says Hilbert: "From the paradise created for us by Cantor, no one will drive us out."

The *intuitionists*, beginning with Kronecker and reinforced by Poincaré, who in our own day are represented by such great minds as Brouwer in Holland, Weyl in Germany, and to a certain extent by Borel in France, have a different story to tell the definition of an aggregate. The disease antedates Cantor, it is deep-seated, and the whole body mathematical is affected. Says Weyl:

> "We must learn a new modesty. We have stormed the heavens, but succeeded only in building fog upon fog, a mist which will not support anybody who earnestly desires to stand upon it. What is valid seems so insignificant that it may be seriously doubted whether analysis is at all possible."

To the intuitionists the issues go far beyond the confines of the theory of aggregates. They maintain that in order that a concept may gain admission into the realm of mathematics, it is not enough that it be "well defined": it must be *constructible*. Not merely must the concept exist in name, but also an actual construction should be given to determine the object which the concept represents. As to construction, the only admissible ones are the finite processes, or—and this is, indeed, a compromise—such infinite processes as are reducible to finite by means of a finite number of rules. The act of conceiving simultaneously an infinite number of single objects and of treating the whole as an individual object does not belong to this category of admissible concepts and must *a priori* be barred from arithmetic. And not only does this mean the scrapping of the theory of aggregates, but even the concept of irrational numbers must undergo a profound modification until analysis is purged of all the impurities with which the indiscriminate use of the infinite has polluted it. "For," says Weyl, "mathematics, even to the logical forms in which it moves, is entirely dependent on the concept of natural number."

While this conflict as to the validity of the foundation on which analysis rests is in full blast, the structure itself is rising at a prodigious rate. Each year sees advances which in the nineteenth century would have required the work of decades. Every decade witnesses the opening of new fields of inductive knowledge which voluntarily submits to the penetration of mathematical analysis. And as to physics, which was among the first conquests of analysis, the pancosmos of the Relativity theory is but a universe of differential forms, and the discontinuous phenomena of the microcosmos seem to obey the laws of a wave mechanics which to all appearances is just an application of a theory in differential equations.

And we see that strange spectacle of men who are loudest in proclaiming that the empire rests on insecure foundations—we see these gloomy deans forsaking from time to time their own counsels of alarum to join in the feverish activity of extending the empire, of pushing further and further the far-flung battle-line.

Of such is the kingdom of logic!

From the day on which man miraculously conceived that a brace of pheasants and a couple of days were both instances of the number two, to this day, when man has attempted to express in numbers his own power of abstraction—it was a long, laborious road, and many were the twists and turns.

Have we reached an impasse? Must we retrace our steps? Or is the present crisis just another of these sharp turns from which, if the future be judged by the past, number will again emerge triumphant, ready to climb to still dizzier heights of abstraction?

CHAPTER 12

The Two Realities

"We have found a strange footprint on the shores of the unknown. We have devised profound theories, one after another, to account for its origin. At last, we have succeeded in reconstructing the creature that made the footprint. And lo! it is our own."

—A. S. Eddington

I have come to the end of my narrative. It was my object to survey the present status of the science of number in the light of its past; so it would be proper in the concluding chapter of such a survey to take a glimpse into the future. But the future belongs to the prophets and I shall respect their prerogative.

There remains the *ever-present*: the issue of reality. This issue has been in the custody of the philosopher since the days when man first consciously attempted to estimate his place in the universe; it is the philosopher's chief preoccupation today.

And so I realize fully that by selecting reality as the theme of this concluding chapter, I am encroaching on a field foreign to my training, foreign to my outlook. I concede further that I have nothing to contribute to the old dilemma; nor have I any intention of rehashing what has been said on the issue by philosophers of opposing schools since the days of Socrates.

My interest lies exclusively in the position which the science of number occupies with respect to the general body of human knowledge. It is from this viewpoint that I shall consider the relation of the number concept to the reality of our senses, in the hope that this may throw light on the historical rôle mathematics has played in creating the *new reality*, the *trans-reality* of modern science.

Between the philosopher's attitude towards the issue of reality and that of the mathematician there is this essential difference: for the philosopher the issue is paramount: the mathematician's love for reality is purely platonic.

The mathematician is only too willing to admit that he is dealing exclusively with acts of the mind. To be sure, he is aware that the ingenious artifices which form his stock in trade had their genesis in the sense impressions which he identifies with crude reality, and he is not surprised to find that at times these artifices fit quite neatly the reality in which they were born. But this neatness the mathematician refuses to recognize as a criterion of his achievement: the value of the beings which spring from his creative imagination shall not be measured by the scope of their application to physical reality. No! Mathematical achievement shall be measured by standards which are peculiar to mathematics. These standards are independent of the crude reality of our senses. They are: freedom from logical contradictions, the generality of the laws governing the created form, the kinship which exists between this new form and those that have preceded it.

The mathematician may be compared to a designer of garments, who is utterly oblivious of the creatures whom his garments may fit. To be sure, his art originated in the necessity for clothing such creatures, but this was long ago; to this day a shape will occasionally appear which will fit into the garment as if the garment had been made for it. Then there is no end of surprise and of delight!

There have been quite a few such delightful surprises. The conic sections, invented in an attempt to solve the problem of doubling the altar of an oracle, ended by becoming the orbits followed by the planets in their courses about the sun. The imaginary magnitudes invented by Cardan and Bombelli describe in some strange way the characteristic features of alternating currents. The absolute differential calculus, which originated as a fantasy of Riemann, became the mathematical vehicle for the theory of Relativity. And the matrices which were a complete abstraction in the days of Cayley and Sylvester appear admirably adapted to the exotic situation exhibited by the quantum theory of the atom.

Yet delightful though these surprises may be, their discovery is not the moving force behind the creative work of the mathematician. For him, mathematics is the field in which he can best manifest his personality. Mathematics for mathematics' sake! "People have been shocked by this formula," said Poincaré, "and yet it is as good as life for life's sake, if life is but misery."

Religion is the mother of the sciences. When the children grew up they left their mother; philosophy stayed at home to comfort the lady in her old age. The long association told more on the daughter than on the mother.

To this day the central problems of philosophy smack of theology. It seems to me that what philosophy lacks most is a principle of relativity.

A principle of relativity is just a code of limitations: it defines the boundaries wherein a discipline shall move and frankly admits that there is no way of ascertaining whether a certain body of facts is the manifestation of the *observata,* or the hallucination of the *observer.*

A principle of relativity is an act of resignation, and a philosophical principle of relativity would consist in the frank admission of the insolubility of the old dilemma: has the universe an existence *per se* or does it exist only in the mind of man? To the man of science, the acceptance of the one hypothesis or the other is not at all a question of "to be or not to be"; for from the standpoint of logic either hypothesis is tenable, and from the standpoint of experience neither is demonstrable. So the choice will forever remain a matter of expediency and convenience. The man of science will act *as if* this world were an absolute whole controlled by laws independent of his own thoughts or acts; but whenever he discovers a law of striking simplicity or one of sweeping universality or one which points to a perfect harmony in the cosmos, he will be wise to wonder what rôle his mind has played in the discovery, and whether the beautiful image he sees in the pool of eternity reveals the nature of this eternity, or is but a reflection of his own mind.

The philosopher's speculations on reality are of little use to us when we attempt to determine the degree of reality we should attach to the general number concept. Other ways must be found, so much is certain. But first let me dispose of certain ambiguities in terms.

The terms used by the mathematician are, after all, words and belong to the limited vocabulary by means of which man from the earliest days had endeavored to express his thoughts, both mathematical and non-mathematical. Some of these terms, such as geometry and calculus, have lost their original double meaning and are understood by everybody in the specific sense that they have acquired in mathematical practice. Others, however, such as logical and illogical, rational and irrational, finite and infinite, real and imaginary, have to this day retained their multiple meaning. To the mathematician, who rarely ventures into the realm of metaphysics, these words have a very specific and quite unambiguous meaning; to the philosopher who uses these terms as his stock in trade they have also a very specific but an entirely different meaning; to the man who is neither philosopher nor mathematician these words have a general and rather vague significance.

No difficulty arises until the philosopher makes an attempt to present to the lay public his analysis of the fundamental concepts of mathematics. It is then that the different connotations attaching to such words as infinity or reality lead to hopeless confusion the mind of the layman.

This particularly applies to the concepts of *real* and *imaginary*. We owe these unfortunate and yet historically unavoidable terms to a philosopher: Descartes. The term *imaginary* as applied to the form $a + \sqrt{-b}$ was justifiable then, inasmuch as no concrete substratum could be assigned to these magnitudes. The moment an interpretation of these magnitudes was found, the inadequacy of the term *imaginary* was realized. This was voiced by Gauss as follows:

"That the subject has been treated from such an erroneous point of view and enveloped with such mysterious obscurity is due largely to the inadequate terminology used. If instead of calling $+1$, -1, $\sqrt{-1}$ the positive, negative and imaginary (sometimes even impossible) unities, they had been called, say, the direct, indirect and lateral unities, this obscurity would have been avoided."

But the protests were largely in vain: the word imaginary had taken deep root. This stability of mathematical terms is phenomenal: it may be due to the conservatism of the mathematician or to his indifference to the choice of words as long as no ambiguity is implied. Be this as it may, eventually the term complex was half-heartedly substituted for imaginary, but to this day both terms are still in use, and as to the word *real,* a change to a more adequate term has not even been proposed.

The use of the term imaginary in this sense is construed by some experts in reality as evidence of the mysticism with which modern mathematics is permeated. They contend that the mathematicians, in selecting these terms, *ipso facto* admitted the fictitious nature of these magnitudes. To argue thus is about as reasonable as it would be for a mineralogist to speak of the stony character of infinitesimal analysis, because the word *calculus* signifies a pebble.

If there be any unreality in the complex number it resides not in the name nor in the use of the symbol $\sqrt{-1}$: a complex number is just a pair of real numbers regarded as a single individual, and it cannot be more fictitious, or less for that matter, than the real numbers of which it is composed. Therefore the critique of the reality of the number concept should revert to the real number. Here the philosopher may find ample evidence of that mysticism which he seeks in mathematics.

However great be the abstraction to which we owe our notion of natural number, the concept was born in the firm "reality" of finite collections. It is true that the moment we began to regard these numbers as a totality, we had to bring in the word *all* with all its implications. Nevertheless the concept of infinity as used in rational arithmetic has been confined to the allegation that *any* number has a successor. The unlimited character of the counting process was invoked only in order that the rules of operation on integers might attain their absolute generality: the infinite was used only as something *potential,* never as an *actuality.*

The rational number is but a pair of integers and therefore has as much reality as the integer. Had we then avoided, as Kronecker urged us to avoid, the introduction of infinite processes and consequently that of the irrationals, the complex number would be just a pair of rational numbers, and whatever reality or unreality we could ascribe to the rational would also reside in the complex. But in the search for a field in which any equation of algebra would have a solution, we were compelled to legitimize the infinite process, and the so-called real number was the result. We do not confine ourselves any more to using infinity as a figure of speech, or as shorthand for the statement that no matter how great a number there is one greater: *the act of becoming* invokes the infinite as the generating principle for any number; any number is now regarded as the ultra-ultimate step of an infinite process; the concept of infinity has been woven into the very fabric of our generalized number concept.

The domain of natural numbers rested on the assumption that the *operation of adding one can be repeated* indefinitely, and it was expressly stipulated that *never shall the ultra-ultimate step of this process be itself regarded as a number.* The generalization to real numbers not only extended the validity of indefinite repetition to any rational operation; it actually abandoned the restriction and admitted the limits of these processes as *bona fide* numbers.

And such is the irony of words, that the so-called *real* numbers have been attained at the sacrifice of a part of that *reality* which we attribute to the natural numbers.

How real are these infinite processes which endow our arithmetic with this absolute generality, which make it the instrument of our geometrical and mechanical intuition, and through geometry and mechanics permit us to express by number the phenomena of physics and chemistry? Well, if reality be restricted to the immediate experience of our senses, no thinking man, be he mathematician, philosopher or layman, would attribute reality to the concept.

There is, however, the widespread opinion that the validity of the infinite is the inevitable consequence of the progress of the empirical sciences. It would be presumptuous on my part to refute this contention in my own words, when David Hilbert has so eloquently answered it in his famous address in memory of Weierstrass:

"The infinite! No other question has ever moved so profoundly the spirit of man; no other idea has so fruitfully stimulated his intellect; yet no other concept stands in greater need of clarification than that of the infinite. ...

"When we turn to the question, what is the essence of the infinite, we must first give ourselves an account as to the meaning the infinite has for reality: let us then see what physics teaches us about it.

"The first naïve impression of nature and matter is that of continuity. Be it a piece of metal or a fluid volume, we cannot escape the conviction that it is divisible into infinity, and that any of its parts, however small, will have the properties of the whole. But wherever the method of investigation into the physics of matter has been carried sufficiently far, we have invariably struck a limit of divisibility, and this was not due to a lack of experimental refinement but resided in the very nature of the phenomenon. One can indeed regard this emancipation from the infinite as a tendency of modern science and substitute for the old adage *natura non facit saltus* its opposite: Nature does make jumps ...

"It is well known that matter consists of small particles, the atoms, and that the macrocosmic phenomena are but manifestations of combinations and interactions among these atoms. But physics did not stop there: at the end of the last century it discovered atomic electricity of a still stranger behavior. Although up to then it had been held that electricity was a fluid and acted as a kind of continuous eye, it became clear then that electricity, too, is built up of positive and negative *electrons*.

"Now besides matter and electricity there exists in physics another reality, for which the law of conservation holds; namely energy. But even energy, it was found, does not admit of simple and unlimited divisibility. Planck discovered the *energy-quanta*.

"And the verdict is that nowhere in reality does there exist a homogeneous continuum in which unlimited divisibility is possible, in which the infinitely small can be realized. The infinite divisibility of a continuum is an operation which exists in thought only, is just an idea, an idea which is refuted by our observations of nature, as well as by physical and chemical experiments.

"The second place in which we encounter the problem of the infinite in nature is when we regard the universe as a whole. Let us then examine the extension of this universe to ascertain whether there exists there an infinitely great. The opinion that the world was infinite was a dominant idea for a long time. Up to Kant and even afterward, few expressed any doubt in the infinitude of the universe.

"Here too modern science, particularly astronomy, raised the issue anew and endeavored to decide it not by means of inadequate metaphysical speculations, but on grounds which rest on experience and on the application of the laws of nature. There arose weighty objections against the infinitude of the universe. It is Euclidean geometry that leads to infinite space as a necessity. … Einstein showed that Euclidean geometry must be given up. He considered this cosmological question too from the standpoint of his gravitational theory and demonstrated the possibility of a finite world; and all the results discovered by the astronomers are consistent with this hypothesis of an elliptic universe."

So the further we progress in our knowledge of the physical world, or in other words the further we extend our perceptual world by means of scientific instruments, the more we find our concept of infinity incompatible with this physical world in deed as well as in principle.

Since then the conception of infinity is not a logical necessity and since, far from being sanctified by experience, all experience protests its falsity, it would seem that the application of the infinite to mathematics *must be condemned in the name of reality*. Such a condemnation would reduce mathematics to the bounded arithmetic and the bounded geometry which I discussed in the fourth chapter. "What is valid seems so insignificant that it may be seriously doubted whether analysis is at all possible." The lofty structure erected by the mathematicians of the last three centuries would be razed to the foundation; the principles and methods which derived their power from the use of the infinite would be scrapped; the physical sciences which have so confidently applied the concepts of limit and function and number in formulating and analyzing their problems would turn over a new leaf: they would rebuild their foundations and devise new instruments in lieu of those condemned.

All this in the name of reality!

It is a drastic program, to be sure. But after this revision has been effected, what little remained of mathematics after this purging process has been consummated would be in perfect consonance with reality.

Would it? That is the question, and this question is tantamount to another: "What is reality?" In asking this latter we are not at all concerned with hair-splitting definitions or with quibbling about irrelevant issues; we are concerned only with the scope we should assign to this reality, which from now on must serve as the criterion as to what is valid and what is not valid.

We naturally turn to the experts in reality. Each one is offering us his particular brand of reality, but as to *the* reality there seems to be no such thing. We are in the state which the French so expressively describe as *embarras du choix.*

Two of these brands are of particular interest to us: the subjective reality and the objectivity reality. The subjective reality seems to be what could be described as the aggregate of all the sense-impressions of an individual. As to objective reality, the definition varies with each philosophical school, for it is precisely here that the dilemma of the existence or the non-existence of a world outside of our conciousness is brought to a climax. Stripped of all its metaphysical irrelevancies and free of philosophical jargon is this description by Poincaré: "What we call objective reality is, in the last analysis, what is common to many thinking beings and could be common to all." In spite of its vagueness, in spite of the obvious weakness of the phrase "what could be common to all," this is the nearest we can get to this intuitive idea of reality which we all seem to possess.

Now the difficulty in determining a valid scope of reality lies precisely in the fact that no individual can successfully separate this subjective reality, which is the aggregate of his personal sense-impressions, from the objective reality which he has acquired from contact with other individuals, present or past. Studies in the psychology of primitive peoples may throw considerable light on this question, but here too environment did its work. The nearest we can come to grasping this subjective reality is the psychology of an infant; and, since we cannot successfully reconstruct our own impressions as infants, we must rely on the studies which adults have made on infants, studies which are invariably colored by preconceived notions.

But let us assume that the subjective reality of an individual can be identified with the data which physiologists and psychologists of the type of Helmholtz or Mach have obtained with regard to the perceptions of sight, sound, touch, etc. If the scope of this reality be taken as the criterion of validity, the unavoidable verdict would be that even the emaciated arithmetic which was the residue left after purging mathematics from the infinite must undergo further amputations; *for the counting process is not a part of this reality.*

The counting process presupposes a different reality, an objective reality, this term being interpreted in the sense Poincaré used it. Counting presupposes the human ability to classify various perceptions under the same head and to endow the class with a name; it presupposes the ability to match two collections, element for element, and to associate these collections with a number-word, which is but the model for a given plurality; it presupposes the ability to order these models into a sequence and to evolve a syntax which will permit an indefinite extension of these number-words. In short, the counting process postulates the existence of a *language*, an institution which transcends the subjective reality or the immediate perceptions of any individual.

If then this subjective reality be taken as criterion of what is valid in mathematics, we should be compelled not only to condemn the infinite process and all it implies, but to scrap the counting procedure as well. The primitive sense, such as some birds and insects possess, would be the only legitimate field for number; and along with our language and arithmetic we should have to scrap the whole intricate structure of civilization which has been erected on these two human institutions.

Of the absolute and immutable world which exists outside our consciousness we know only through theological speculations: accepting it or rejecting it are alike futile to a natural philosophy. But just as futile and sterile is the acceptance of the crude reality of our senses as the *arch-reality*, the only reality. It is convenient for a systematic exposition, to be sure, to regard the newly born child, or the primitive man, or the animal, as the embodiment of such an arch-reality. We can go still further and imagine, as did Helmholtz, Mach and Poincaré, an intelligent being who has been deprived of all but one of his senses, say sight, and speculate on the type of universe that such a being would construct. Such speculations are tremendously fascinating in that they allow free rein to our power of resolving our sensations into their constituents, and then regarding the concept as a synthesis of these *arch-sensations*. But to accept such a synthesis as reality, as *the* reality, has, to my way of thinking, one fatal defect: it postulates the existence of an individual intellect; whereas the very process of coördinating these sensations involves thought, which is impossible without the vehicle language, which in turn implies an organized exchange of impressions, which in turn presupposes a collective existence for human beings, some form of social organization.

The only reality that can be taken as a criterion of validity is not that absolute, immutable reality which exists outside of our consciousness and is therefore pure metaphysics, nor that arch-reality which the physiologist and the psychologist manage to isolate by means of painstaking experiments; it is rather that objective reality which is common to many and could be common to all. And that reality is not a collection of frozen images, but a living, growing organism.

But when we turn to this objective world with a view to finding there a criterion for the reality of mathematical concepts, we strike a new difficulty. These things common to many may be confined to those immediate impressions of an individual which are shared by other thinking beings; but they may also include all the data which the race has acquired through the application of scientific instruments, for such facts, too, are common to many and could presumably be common to all. Now this latter extended world may be valid for judging the reality of any qualitative statement, but the moment we attempt to use it as a criterion for number, we are confronted with the fact that this objective world presupposes number, because our scientific instruments are designed, built, and used according to fixed mathematical principles which, in turn, rest on number.

Indeed, whether we use a ruler or a weighing balance, a pressure gauge or a thermometer, a compass or a voltmeter, we are always measuring what appears to us to be a *continuum*, and we are measuring it by means of a graduated *number scale*. We are then assuming that there exists a perfect correspondence between the possible states within this continuum and the aggregate of numbers at our disposal; we are tacitly admitting an axiom which plays within this continuum the rôle which the Dedekind-Cantor axiom plays for the straight line. Therefore, any measuring device, however simple and natural it may appear to us, implies the whole apparatus of the arithmetic of real numbers: behind any scientific instrument there is the master-instrument, arithmetic, without which the special device can neither be used nor even conceived.

This is then the difficulty: if we were to judge the reality of the real number by the objective world which contains all the data derived by means of scientific instruments, we should be moving in a vicious circle, because *these instruments already postulate the reality of the real number.*

Nor would we avoid the vicious circle if we were to confine ourselves to the more restricted objective world of those of our immediate impressions that are shared by others. For let us ban all measuring devices, let us declare public opinion the sole criterion of reality. How shall we then arrive at a valid judgment? When you declare that I am color-blind, because I see green where you see red, how will you demonstrate the truth of your assertion, if not by an appeal to *majority rule?* Here too then we have both agreed to leave the *decision to number.*

Figures do not lie, because they cannot lie. They cannot lie, because they have been declared *a priori* infallible. Having elected number as the sole arbiter for judging values, having agreed to abide by its decisions, we have *ipso facto* waived our right of appeal to any other tribunal.

What is the conclusion?

An individual without a milieu, deprived of language, deprived of all opportunity to exchange impressions with his peers, could not construct a science of number. To his perceptual world arithmetic would have no reality, no meaning.

On the other hand, the objective world of a thinking being is made up of those impressions which are shared by the majority of his peers. To him the question: what reality shall we ascribe to number? is meaningless, because there is no reality without number, as there is no reality without space or without time.

And so neither in the subjective nor yet in the objective world can we find a criterion for the reality of the number concept, because the first contains no such concept, and the second contains nothing that is free of the concept.

How then can we arrive at a criterion? Not by evidence, for the dice of evidence are loaded. Not by logic, for logic has no existence independent of mathematics: it is only one phase of this multiphased necessity that we call mathematics. How then shall mathematical concepts be judged? *They shall not be judged!* Mathematics is the supreme judge; from its decisions there is no appeal.

We cannot change the rules of the game, we cannot ascertain whether the game is fair. We can only study the player at his game; not, however, with the detached attitude of a bystander, for we are watching our own minds at play.

I recall my own emotions: I had just been initiated into the mysteries of the complex number. I remember my bewilderment: here were magnitudes patently impossible and yet susceptible of manipulations which lead to concrete results. It was a feeling of dissatisfaction, of restlessness, a desire to fill these illusory creatures, these empty symbols, with substance. Then I was taught to interpret these beings in a concrete geometrical way. There came then an immediate feeling of relief, as though I had solved an enigma, as though a ghost which had been causing me apprehension turned out to be no ghost at all, but a familiar part of my environment.

I have since had many opportunities to realize that my emotions were shared by many other people. Why this feeling of relief? We have found a concrete model for these symbols; we have found that we can attach them to something familiar, to something that is real, or at least seems real. But why consider a point in the plane—or rather the segments measuring the distances of such a point from two arbitrary axes of references—why consider this more real than the quantity $a + ib$? What reality is there behind a plane, a line, a point? Why, only a year or two before, these too were but phantoms to me. The plane extending indefinitely in all directions—my nearest approximation of this was a sheet of paper, 8 inches by 11, or a wavy blackboard, full of ridges and scratches. The line deprived of thickness; the point of intersection of two such lines—a thing without dimension, a pure illusion, for which there existed no model whatsoever; and finally the coördinates of such a point, which involved all the uncertainties, all the inaccuracies of measurement—were such things the concrete reality that caused my feeling of relief?

We have attached a phantom to a fiction, which had this advantage over the phantom that it was a *familiar fiction.* But it had not always been familiar; there was a time when this too caused bewilderment and restlessness, until we attached it to a still more primeval illusion, which, in turn, had been rendered concrete through centuries of habit.

The reality of today was but an illusion yesterday. The illusion survived because it helped to organize and systematize and guide our experience and therefore was useful to the life of the race. Such is my interpretation of the words of Nietzsche:

> "We hold mere falsity no ground for rejecting a judgment. The issue is: to what extent has the conception preserved and furthered the life of the race? The falsest conceptions,—and to these belong our synthetic judgments *a priori,*—are also those which are the most indispensable. Without his logical fictions, without measuring reality in a fictitious absolute and immutable world,

without the perpetual counterfeiting of the universe by number,
man could not continue to live. The renunciation of all false
judgment would mean a renunciation, a negation of life."

It is not the immediate evidence, nor the laws of logic, that can
determine the validity of a mathematical concept. The issue is:
how far does the concept preserve and further the intellectual
life of the race? That is why the alarming reports of the gloomy
deans leave me unalarmed. The criterion of validity for any illusion is a *post factum* and sometimes a *post mortem* judgment.
Those that preserve and further the life of the race thrive and
grow, and thus earn their right to reality; those that are harmful
or useless eventually find their way to the textbooks on metaphysics and theology, and there they stay. So they, too, do not die
in vain.

Experimental evidence and logical necessity do not exhaust the
objective world which we call reality. There is a mathematical
necessity which guides observation and experiment, and of
which logic is only one phase. The other phase is that intangible,
vague thing which escapes all definition, and is called intuition.
And so, to return to the fundamental issue of the science of
number: the infinite. The concept of infinity is not an experien-
tial nor a logical necessity; it is a *mathematical necessity.* This
affirmation of the power of the mind which knows itself capable
of conceiving *the indefinite repetition of an act, when this act is at
all possible, may be pure fiction, but it is a convenient and therefore a necessary fiction.* It frees us from the burden of examining
in each particular case whether what we assert about the case is
at all possible. It lends to our statement that appearance of generality without which there would be no science; and above all it

bridges the chasm between the inescapable conception of a world which *flows with the stream of time,* and the number concept which was born in *counting the discrete.*

But the infinite is only one of the many by-paths through which the Quest of the Absolute has taken man. There are many others: simplicity, uniformity, homogeneity, regularity, causality are other manifestations of this mathematical intuition. For it is mathematical intuition that urges the mind on to follow the mirage of the absolute and so enriches the intellectual heritage of the race; but when further pursuit of the mirage would endanger this heritage, it is mathematical intuition that halts the mind in its flight, while it whispers slyly: "How strangely the pursued resembles the pursuer!"

And what is the source of this creative intuition? What is this necessity which organizes and guides human experience and shields it from the terrors of Chaos? Whence came this thrust that raised the frigid, immobile, and barren rocks of logic?

> "The billows are whispering their eternal whispers,
>   The wind blows on, the clouds are sailing,
>   The stars keep twinkling indifferent and cold,
>   And a fool waits for his answer."

And the wise man? The wise man, resuming his business in life, the spinning of the fictions of today, which may be the realities of tomorrow, casts one last glance at the distant peaks, behind which is lost the origin of thought, and repeats the words of the Master:

> "THOUGH THE SOURCE BE OBSCURE,
> STILL THE STREAM FLOWS ON.

Appendixes

On the Recording of Numbers

*"For, contrary to the unreasoned opinion of the igno-
rant, the choice of a system of numeration is a mere
matter of convention."*

—Pascal

On the Number Sense of Beast and Man

How can one discern that the number of objects in a collection has changed, without having recourse to actual counting? What is the nature of this intuition which we call *Number Sense?* Quite a few readers of the earlier editions of this work have attempted to answer these questions. I do not consider myself competent to pass on the validity of the arguments adduced in support of their theories, but neither do I want to impose restrictions on their creative imagination, and so I shall list here a few of the more cogent conjectures of these correspondents.

The *heterogeneity* of a collection may assist in the estimate. Say that on entering a room you have recognized that there are fewer people than usual, because a familiar face is missing: you owe the success of your estimate to the circumstance that the members of the group were not like "peas in a pod," but individuals, each with distinctive traits of his own. In some such way, perhaps, did the crow in the story of Chapter One discern that not all who had entered the tower came out of it.

261

Fatigue resulting from an effort to overcome an obstacle may facilitate the appraisal. Thus, if you have ascended a staircase without counting the number of floors, your limbs may tell whether you have climbed five or six flights. The number prowess of the solitary wasp may be explained on some such grounds.

Quite often *pattern reading* is of considerable help. If you have recognized at a glance that a table has been set for more persons than the customary four, it is the changed pattern that has conveyed to you this information. Or, consider a row of peas on a table: if they were arranged close enough together and "in line," you might not be able to tell whether there were five peas or six; you would be more likely to guess right if the distribution were non-uniform, and you would trust your judgment even more if the peas were arranged as vertices of a polygon. Some such principle may govern the instinct of a bird whose nest has been robbed.

How Xerxes Counted His Army

"The territory of Doriscus is in Thrace, a wide plain by the sea, and through it flows a great river, the Hebrus; here was built that royal fortress which is called Doriscus, and a Persian guard was posted there by Darius ever since the time of his march against Scythia. It seemed, therefore, to Xerxes to be a fit place to array and number his host, and this he did. All the fleet, having now arrived at Doriscus, was brought at his command to a neighboring beach ... and hauled up for rest. ...In the meanwhile Xerxes numbered his army....

"What the number of each part was I cannot with exactness say, for there is no one who tells us that; but the count of the whole land army showed it to be a million and seven hundred thousand. The numbering was done as follows: a myriad men were collected in one place, and when they were packed together as closely as might be, a line was drawn around them; this being

done, the myriad was sent away, and a wall of stone built on the line reaching up to a man's navel; which done, others were brought into the walled space, till in this way all were counted."
Herodotus: *Historia*, Book VII.

On Recording Large Numbers

"Hence it is evident that the number of grains of sand contained in a sphere as large as the one bounded by the fixed stars, the diameter of which has been estimated by Aristarchus, is less than one thousand myriads of units of the eighth class." Thus concludes Archimedes his tract known as the "Sandreckoner."

What did he mean by units of the eighth class? The Greek *myriad* stood for ten thousand; this number, $M = 10^4$, Archimedes designated as *unit of the first class.* Next came the *octad*, i.e. a myriad myriads, which Archimedes defined as *unit of the second class.* Let us denote the octad by Ω: then $\Omega = M^2 = 10^8$. We would next infer that $M^3 = 10^{12}$ is the unit of the third class; however, such was not the case. The *base* of the Archimedean scheme was the *octad* $\Omega = 10^8$. Thus, *unit of class n* should be interpreted as Ω^{n-1}, and the "number of grains in the universe," as estimated by Archimedes, was

$$10^3 \times 10^4 \times (10^8)^7 = 10^{63}$$

Expressed in modern terms this estimate would read: the number of grains ... is of *magnitude* 10^{63}.

It is instructive to compare this Archimedean estimate with the largest prime "known," which is the *seventeenth Mersenne number* recently calculated by the Institute of Numerical Analysis at Los Angeles:

(1) $$M = 2^{2281} - 1$$

This integer is of the magnitude 10^{687}, and since

$$687 = 85 \times 8 + 7,$$

Archimedes would have described it as of the order of *one thousand myriads of the eighty-sixth class.*

On the Principle of Position

Most of us have learned to count at a very tender age, and few of us have had occasion to reflect on the subject since. Let us refresh our memories.

It all begins as a sort of coordination between *fingers* and *lips.* The child learns to associate certain patterns formed by his fingers or blocks with certain words. He is told that these words are called numbers, and he is made to memorize them in an *ordered series.* By the time he has run short of fingers and blocks, he is taught to use a quaint *rhetorical* procedure which enables him to extend his counting scope without recourse to new tangible patterns. By now, counting has turned into a race with numbers or a game of words, a game which the child is only too eager to play, at first. Until one day he comes to realize that, after all, *what has been said or done once can always be repeated*; then stopping abruptly at some term in this number series, he dismisses the rest with an impatient "and so forth and so on." On that day his education in counting has been completed. There is planted, at the same time, in his mind the germ of an idea which many years later will rise to perplex him in the guise of the *concept of infinity.*

What is the rhetorical procedure which induces this supreme confidence that the counting process can be carried on indefinitely? Well, in spite of the fact that we have mastered the procedure at so tender an age, it rests on a rather intricate mathematical idea. This idea posits that *any positive integer may be represented in one and only one way as a polynomial arranged in powers of ten, the coefficients of the polynomial being restricted to integers less than ten.*

As was pointed out in Chapter Two, the preference for the base *ten* is not based on any intrinsic merits of that integer, but is patently the consequence of the *physiological* accident that most normal people have ten fingers on their hands. This physiological aspect of our number language is curious enough; but even more striking is its *polynomial structure*. It appears indeed that wherever man had been forced by the vicissitudes of life to deal in integers larger than what his finger technique could handle, he invariably resorted to this polynomial representation.

Positional numeration is but an adaptation of this rhetorical procedure to writing; by adjoining to the range of admissible coefficients the symbol 0, we fill such "gaps" in the terms of the ordered polynomial as may occur, and hence forestall any possible ambiguity. Thus (abcd) is but a *cryptogram*, an abbreviated form of the polynomial:

$$aR^3 + bR^2 + cR + d = (abcd)_R,$$

where R is the *base*, and the coefficients a, b, c, d, can range over the values 0, 1, 2, ..., R − 1. As to R, the base or *radix* of the system, any positive integer other than 1 may be selected. Moreover, since the operations of arithmetic which we have learned to apply to numbers expressed in decimal notation derive from the properties of general polynomials, these rules can be readily adapted to any other system.

On the History of the Decimal Mark

Positional numeration had been in full use for many centuries before it was realized that among the advantages of the method was its great facility in handling fractions. Even then the realization was far from complete, as may be gleaned from the cumbersome superscripts and subscripts used by Stevin and Napier.

Now, all that was necessary to bring the scheme to full effectiveness was a mark such as our modern *decimal point,* separating the integral from the fractional part of the number. Yet, for some unaccountable reason the innovators, with the exception of Kepler and Briggs, either did not recognize this fact, or else had no faith that they could induce the public to accept it. Indeed, a century after Stevin's discovery, a historian of the period, referring to the many conflicting notations then in use, remarked: *Quod homines tot sententiae* (As many opinions as there are people), and it took another century before the decimal notation was finally stabilized and the superfluous symbols dropped.

Author	Time	Notation
Before Simon Stevin		$24\dfrac{375}{1000}$
Simon Stevin	1585	$24\quad 3^{(1)}\ 7^{(2)}\ 5^{(3)}$
Franciscus Vieta	1600	$24 |_{375}$
John Kepler	1616	24(375
John Napier	1617	$24:3\overset{\text{I}}{7}\overset{\text{II}}{}\overset{\text{III}}{5}$
Henry Briggs	1624	24^{375}
William Oughtred	1631	$24\lfloor\underline{375}$
Balam	1653	24 : 375
Ozanam	1691	$24\cdot\overset{(1)}{3}\overset{(2)}{7}\overset{(3)}{5}$
Modern		$24\cdot 375$

On Choice of Base

Buffon's proposal that the universal base of numeration be changed from *ten* to *twelve* enjoyed a curious revival in our own century. Duodecimal Societies sprang up both here and abroad; pamphlets and periodicals appeared extolling the virtues of *twelve* with a zeal akin to religious fervor; having settled the "vexing" question of symbols for *ten* and *eleven,* the reformers turned to tabulation. There were *conversion tables, multiplication tables,* and even tables of *duodecimal logarithms.* Eventually, the crusade faded away, like so many other movements which aim at reforming a collective habit of the race.

Quite different has been the fate of another reform, earlier in vintage and even more exotic than the one proposed by Buffon: the *Binary Arithmetic* of Leibnitz. Alas! What was once hailed as a monument to monotheism ended in the bowels of a robot. For most of the modern high-speed calculating machines operate on the *binary* principle. The robot is not tied to human traditions, it forms no habits and its "memory" is controlled by "coding." The lack of compactness of binary records is richly compensated by the prodigious speed of the machine. Nor are the arithmetical habits of the human operator seriously challenged, inasmuch as the robot will automatically turn decimal data into binary and back. Indeed, the time may not be far off when the man behind the machine will be as little aware of the binary bowels of his robot as he is of his own inner organs.

On Change of Scale

The limited scope of man's number sense makes it well nigh impossible to name a given number after some model-collection of which it is the "cardinal measure." The alternative is to associate the number with the symbols used in recording it. Prior to the introduction of Arabic numerals, letters of an alphabet were used for the purposed, and this may account in part for the great success which Gematria enjoyed in those days. With the advent of positional numeration and its universal acceptance, the decimal cryptogram of a number automatically provided it with a name. Today it has become more than a name: *we have learned to identify the number with its decimal cryptogram.* So great indeed is the force of habit, that most of us regard any other representation of a number as a sort of disguise; and this despite the

fact that we all realize that there is nothing absolute or sacro-
sanct about the scale of ten.

The passage from any scale to decimal notation is suggested
by the defining polynomial of the cryptogram. Take, for example, the cryptogram $(4321)_5$: by definition

$$(4321)_5 = 4 \times 5^3 + 3 \times 5^2 + 2 \times 5 + 1 = 586.$$

The calculation can be made less laborious by an artifice which
the reader has learned under the name of *synthetic division*, but
for which *synthetic substitution* would have been a more fitting
name. The algorithm is carried out in detail in the following
table:

		4	3	2	1
			$(5 \times 4) + 3$ $= 23$	$(5 \times 23) + 2$ $= 117$	$(5 \times 117) + 1$ $= 586$
Quotients		4	23	117	586
Remainders		4	3	2	1

The same table shows how to transcribe a number expressed in
traditional terms to any other scale. By *reversing* the algorithm,
we obtain a chain of successive divisions by the radix of the new
scale, 5 in the example chosen: *the remainders of these divisions
form the digits of the cryptogram sought.*

As an application let us calculate the successive terms
of the so-called *Mersenne sequence,* the general term of which is
$M_p = 2^p - 1$. In *binary* notation these integers are expressed as sets
of units:

(2) $M_1 = 1, M_2 = (11)_2, M_3 = (111)_2, M_4 = (1111)_2, \ldots$

The algorithm described above offers a convenient method
for calculating these integers:

p	1	2	3	4	5	6	7	8	9	10
	1	1	1	1	1	1	1	1	1	1
M_p	1	<u>3</u>	<u>7</u>	15	<u>31</u>	63	<u>127</u>	255	511	1023

The underlined terms are the *primes* of the sequence and are called *Mersenne numbers*. I shall return to these integers in a subsequent article (see B 10).

A Problem of Pascal

The quotation put at the head of this chapter is a passage taken from an essay of Pascal entitled "On the Divisibility of Numbers as Deduced from Their Digits." Pascal's test of divisibility by an integer q is closely related to the expansion of 1/q into a decimal *fraction*. It was from this point of view that Pascal's study was taken up by his British contemporary, John Wallis, and in the following century by John Bernoulli, Euler, and Lambert, who extended and sharpened the theory.

Let us examine the method used in expanding into a decimal fraction the number 1/q, where q is any prime other than 2 or 5. The algorithm for this conversion is the familiar *long division;* the decimal fraction generated in the process is of a type known as *periodic,* because it consists of an infinite number of identical "blocks." The block is called a *cycle,* and the number of digits in the block the *period* of the cycle. Thus, in the case q = 7 we find

Quotients	.0.	1	4	2	8	5	7	1	4	2	8	5	7	...
Dividends	1	0	0	0	0	0	0	0	0	0	0	0	0	...
Remainders	1	3	2	6	4	5	1	3	2	6	4	5	1	...

Here the cycle is K = 142857 and the period is p = 6.

The quotients, which generate the digits of the expansion, have no bearing on Pascal's argument. He is concerned only with the remainders, which, for want of a better name, I shall call the *decimal residues of the divisor q*. The sequence of residues is *periodic*, and the *period of the residue cycle determines the period of the expansion*. It follows from the nature of the algorithm that every residue cycle begins with 1, and that *no two residues in the same cycle can be equal*. On the other hand, the residues may assume any value from 1 to $q - 1$: hence, *the period is at most equal to $q - 1$*. As a matter of fact, the period is either $q - 1$, as is the case of $q = 7$, or some *divisor of $q - 1$*, as is the case of $q = 13$. In the table below are listed the residues of various integers; following Pascal, the listing is *from right to left*, to conform with the order of the digits in a cryptogram.

I shall now present Pascal's argument in the case $q = 7$, which is typical of the general situation. First, let us "spell out" in detail the results of the long division:

$$1 = 0 \cdot 7 \quad +1 \qquad 10^3 = 142 \cdot 7 \quad +6 \qquad 10^6 = 142857 \cdot 7 + 1$$
$$10 = 1 \cdot 7 \quad +3 \qquad 10^4 = 1428 \cdot 7 \quad +4 \qquad \dots\dots\dots\dots\dots$$
$$10^2 = 14 \cdot 7 \quad +2 \qquad 10^5 = 14285 \cdot 7 \quad +5$$

Next consider the three-digit number

$$N = (CBA) = A \cdot 1 + B \cdot 10 + C \cdot 10^2$$

Substituting for the powers of 10 their values, we are led to

$$(3) \qquad\qquad N = 7 \cdot H + (A + 3B + 2C),$$

where H is some positive integer: it follows that N is divisible by 7 if, and only if, 7 divides $A + 3B + 2C$.

Passing now to the general case, let $R_1, R_2, R_3, \dots, R_j, \dots$ be the residues of q, and let $N = (D_j D_{j-1} \dots D_3 D_2 D_1)$ be the number to be tested for its divisibility by q: then N *is or is not divisible by q* according as *q does or does not divide the weighted sum*

Decimal Residues and Cycles

Divisor q	Period P	10^8	10^7	10^6	10^5	10^4	10^3	10^2	10	1
2		0	0	0	0	0	0	0	0	1
3	1	1	1	1	1	1	1	1	1	1
5		0	0	0	0	0	0	0	0	1
7	6	2	3	1	5	4	6	2	3	1
9	1	1	1	1	1	1	1	1	1	1
10		0	0	0	0	0	0	0	0	1
11	2	1	10	1	10	1	10	1	10	1
13	6	9	10	1	4	3	12	9	10	1
17	16	16	5	9	6	4	14	15	10	1

(4) $$P = R_1 D_1 + R_2 D_2 + R_3 D_3 + \dots + R_j D_j$$

We expressed Pascal's theorem in terms of *divisibility*. However, the result is susceptible of a broader interpretation. Indeed, the reasoning reveals that *the remainder of division of a number N by q is equal to the remainder of division by q of the test-function P*, or, to use a terminology first introduced by Gauss, that *the integers N and P are congruent modulo q*. In symbols

(5) $$N \equiv P \pmod{q}$$

Thus in the case of q = 9, we find $R_1 = R_2 = R_3 = \dots = 1$, and, consequently, $P = D_1 + D_2 + D_3 + \dots D_j = \Sigma D$,

(6) $$N \equiv \Sigma D \pmod 9$$

This is the so-called *rule of nine*, which states that *whether we divide the integer or its digital sum by 9, the remainder is the same*. In particular, *a number is divisible by 9 if and only if 9 divides its digital sum*.

On Digits and Divisors

Pascal's test of divisibility of N by q holds for all values of N and q; in practice, however, one soon reaches the point of "diminishing returns." On the other hand, some of the rules which follow from the general theorem are of more than passing interest to the practical computer. Before presenting these rules let me remark that in testing for divisibility by an integer q, one can always replace the Pascal *test-function*, P, by a *congruent* form, or, in simple terms, one can *add* or *subtract* from P any *multiple of q* without affecting the validity of the test. For example, Pascal's criterion for divisibility of a 3-digit number by 7 is that 7 divides $P = 2C + 3B + A$. By adding 7B and setting $10B + A = T$, we get the more convenient test-function: $T + 2C$. Thus 581 is divisible by 7 because 7 divides $81 + 2 \cdot 5 = 91$.

I. *Testing a 3-digit Number:* $N = (CBA)$, $T = (BA)$

Modulus: q	Multiple of q nearest to 100	Test Function
7	98	$2C + T$
11	99	$C + T$
13	104	$T - 4C$
17	102	$T - 2C$
19	95	$5C + T$
23	92	$8C + T$
29	87	$13C + T$
31	93	$7C + T$

Example: $N = 912$, $T = 12$. N is not divisible by 17 because $12 - 2 \cdot 9 = -6$; N is divisible by 19, because $12 + 5 \cdot 9 = 57 = 19 \cdot 3$.

II. *Criterion for II.* The Pascal function is $P = A + 10B + C + 10D + E + 10F + \ldots$ By subtracting $11B + 11D + 11F + \ldots$, we reduce the criterion to the more convenient form: $P' = A - B + C - D + E - F + \ldots$.

Example: $N = 399{,}168$ is divisible by 11, because $3 - 9 + 9 - 1 + 6 - 8 = 0$.

III. *Criteria for 7 and 13.* These, too, can be derived from Pascal's theorem. However, a more direct approach is offered by the circumstance that *1,001 = 10³ + 1 is divisible by both 7 and 13,* and that, consequently, both 7 and 13 divide $10^6 - 1 = 999,999$, $10^9 + 1 = 1,000,000,001$ etc. To be specific, let N be any nine-digit integer; we can write

(7) $N = X + 1,000Y + 1,000,000Z = X - Y + Z + 1,001H$

Thus N is divisible by 7 (or 13) if 7 (or 13) divides $X - Y + Z$. This simple rule is particularly adapted to the modern trend of recording large numbers in blocks of three.

Example: $N = 864,192$. Here $864 - 192 = 672 = 7 \cdot 96$. Hence, N is divisible by 7, but not by 13.

The Residue Check

Nothing is known of the discoverer of the Rule of Nine, or as to how long it has been in use, except that Pascal in his essay on divisors and digits, which I quoted before, refers to it as common knowledge, and that pamphlet is now more than three hundred years old. Accountants of a less sophisticated age used the rule for checking additions and multiplications. I doubt whether contemporary bookkeepers ever indulge in such finesse. For better or for worse, the advent of calculating machines has rendered the computer's skill as obsolete as fine penmanship.

Strangely enough, the Rule of Nine derives from a principle which transcends in scope the rather trivial applications for which it had been originally designed. Indeed, the technique is valid for any modulus and in any system of numeration, and its teaching could serve as an excellent introduction to the Gaussian Theory of Congruences which I mentioned in the preceding article. However, to avoid confusion, I shall use a more direct approach.

If a is the remainder of division of the integer A by the integer q, then I shall say that *a is the residue of A modulo q* and write

(8)   res A = a(mod q)

The residue may range from 0 to q – 1, and res A = 0 (mod q) means that q divides A. Let now A, B, C, ... be any finite set of integers, and let a, b, c, ... be their residues with respect to the same modulus q. Then it is not difficult to prove that

(9) $\begin{cases} \text{res } (A + B + C + ...) = \text{res } (a + b + c + ...) \\ \text{res } (A - B) = \text{res } (a - b) \\ \text{res } (A \cdot B \cdot C \cdot ...) = \text{res } (a \cdot b \cdot c \cdot ...) \\ \text{res } (A^m) = \text{res } (a^m) \\ \text{res } (A^m \cdot B^n \cdot C^p \cdot ...) = \text{res } (a^m \cdot b^n \cdot c^p \cdot ...) \end{cases}$

where the exponents m, n, p, ... are positive integers.

By combining these properties we can extend the proposition to the most general *integral function, bearing on any number of integral arguments, coefficients and parameters.* I shall call this general theorem the *residue principle*; it can be stated as follows: Let F (x, y, z, ...) be any integral function, and suppose that the substitution x = A, y = B, z = C, etc., leads to an integer N; furthermore, let a, b, c, ..., n be the residues of A, B, C, ..., N, with respect to some modulus q; then

(10)   F(A,B,C,...) = N implies F(a,b,c,...) = n

How the residue principle may be used for checking numerical computations will now be shown on a set of examples which have been chosen for their historical interest.

Historical Illustrations

I. As stated by Fermat and proved by Euler, the equation $x^3 + y^3 = R^3$ has *no integral solutions*. On the other hand, the equation,

(11) $$x^3 + y^3 + z^3 = R^3,$$

admits of an *infinite number of solutions*. Some of these, like (3,4,5;6) and (1,6,8;9), were already known to Fibonacci. A list of solutions comprising more than one hundred sets was published in 1920 by H.W. Richmond. One of these sets is (25,38,87;90). To check that

$$25^3 + 38^3 + 87^3 = 90^3$$

observe that the last two terms are divisible by 9: hence 9 must divide $25^3 + 38^3$, and indeed $25 + 38 = 63$. On the other hand, 25^3 and 90^3 contain the common divisor 125; hence 125 must divide $38^3 + 87^3$. As it happens, $38 + 87 = 125$.

II. The same list asserts that

(12) $$243 + 633 + 893 = 98^3.$$

Since 7 divides 98 and 63, it must divide $243 + 893$. Now,

$$\text{res } 24^3 \pmod 7 = \text{res } 3^3 = 6 \text{ and res } 89^3 = 5^3 = 6.$$

Hence,

$$\text{res } (24^3 + 89^3) \pmod 7 \text{ is 5, not 0.}$$

The entry is, therefore, erroneous, and yet the nine-test gives an affirmative check.

III. Closely related to the preceding question is a *problem of Ramanujan*: to determine the integers which may be *partitioned into a sum of two cubes in more than one way*. $1{,}729 = 10^3 + 9^3 = 12^3 + 1^3$ is the smallest integer of this type. Check that

(13) $$N = 1{,}009{,}736 = 96^3 + 50^3 = 93^8 + 59^3.$$

The nine-test yields 8 = 8 = 8. On the other hand, N is divisible by 7 since 7 divides 736 − 9 + 1 = 728. The checks, modulo 7, lead to 126 = 7 · 18 and 35 = 7 · 5.

IV. Check that

$$N = 12! + 1 = 479,001,601.$$

Since 12! is divisible by any integer less or equal to 12, the remainder of division of 479,001,601 by any such integer must be 1, and this the reader can readily verify. An additional check is provided by *Wilson's theorem* which asserts that 12! + 1 is divisible by the *prime 13*. We find indeed that

$$\text{res } (479,001,601) = \text{res } (479 − 1 + 601) = \text{res } (1,079)$$
$$= \text{res } (78) = 0 \ (\text{mod } 13)$$

V. How Euler found out that *1,000,009 is not a prime,* but the product of the primes 293 and 3,143, is told elsewhere (See *"Formuae for Primes,"* in Appendix B). To check that

$$1,000,009 = 293 · 3,413$$

we use, first, the nine-test. The digital sums are 10, 14 and 11; the residues 1, 5, and 2, and res (5 · 2) = 1. Next, the seven-test: res (1,000,009) = res (9+1) = 3; res (293) = res (93 + 4) = 6; res (3,413) = res (410) = res (10 + 8) = 4; and res (4 · 6) = 3.

VI. No prime factor of a 3-digit number can exceed 31, since $31^2 < 1,000 < 37^2$. Hence, the table of test-functions of Article 9 can be conveniently used to check the *primality of any 3-digit number*. Thus I mentioned on page 51 that the *fifth Fermat number* $2^{2^5} + 1$ *was divisible by 641*. Is 641 a prime? We have here C = 6, T = 41, and, since $29^2 > 641$, the primes to be tested are: 7, 11, 13, 17, 19, and 23. We find that none of the corresponding residues is zero and conclude that 641 is a prime.

Topics in Integers

"We have found a beautiful and most general proposition, namely, that every integer is either square, or the sum of two, three or at most four squares. This theorem depends on some of the most recondite mysteries of number, and it is not possible to present its proof on the margin of this page."

—Fermat

Two Arithmetic Triangles

The use of patterns for bringing out properties of integers did not die with the Pythagoreans. Pascal's *arithmetic triangle* is a case in point. Less familiar, but fully as elegant is Fibonacci's proof of the identity

(14) $$1^3 + 2^3 + 3^3 + \ldots + n^3 = (1 + 2 + 3 + \ldots + n)^2$$

Having arranged the consecutive odd integers in a *triangular array*, as shown in Figure 1, Fibonacci observed that the k terms of the k-th row form an *arithmetic progression of mean value* k^2; hence the sum of the terms of the k-th row is $k \times k^2$ or k^3, and the sum of all terms in n consecutive rows is $S = 1^3 + 2^3 + 3^3 + \ldots + n^3$. On the other hand, in virtue of a proposition which tradition attributes to Pythagoras himself, the *sum of the first p odd integers is equal to p^2*; thus

$$S = (1 + 2 + 3 + \ldots + n)^2.$$

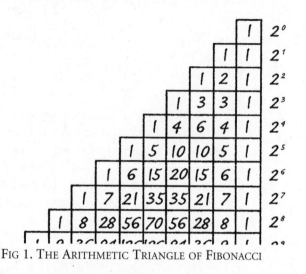

FIG 1. THE ARITHMETIC TRIANGLE OF FIBONACCI

Pascal's triangle was designed to bring out the relation between *binomial coefficients* of consecutive orders. Let us agree to denote by (α,β) the coefficient of the term $x^\alpha y^\beta$. The expansion $(x + y)^p$ contains this term if $\alpha + \beta = p$; it also contains the term $x^{\alpha-1}y^{\beta+1}$. On the other hand, the *binomial expansion*, $(x + y)^{p+1}$, contains the term $x^\alpha y^{\beta+1}$. Between the coefficients of these "contiguous" terms exists the relation

(15) $(\alpha,\beta) + (\alpha - 1, \beta + 1) = (\alpha,\beta + 1)$

This is *Pascal's Recursion Law.*

Now, if it be true, as has been often asserted, that it was the contemplation of this "mystic figure" that had led Pascal to fomulate the *principle of mathematical induction,* then the Arithmetic Triangle should be enshrined in the Museum of

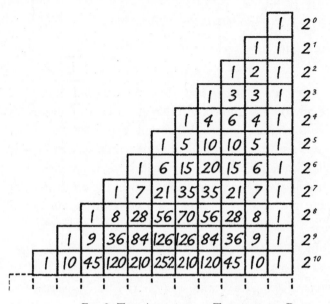

FIG 2. THE ARITHMETIC TRIANGLE OF PASCAL

Mathematical History. And yet, as a technical device it exerted little influence on subsequent developments in this field. This was partly due to the limitations of the method, which was not susceptible of ready generalizations either toward multinomial expansions or toward negative and fractional exponents. Also, the very form of the Pascal recursion law obscured the rather important connection between *binomial* and *factorial* integers. But the chief cause of this failure is to be found in the history of the period which followed the great triumvirate: Descartes, Pascal, Fermat. The emergence of infinitesimal analysis overcast the brilliant achievements of those men, and the Theory of Numbers suffered most from this eclipse.

The Multinomial Theorem

The expression

$$(x + y + z + \ldots + w)^p$$

where p is a positive integer, and where the n *arguments,* x, y, z, … w, are any entities subject to the general laws of algebra, is called a *multinomial of order p.* When fully expanded, the multinomial gives rise to a polynomial of degree p. This polynomial is *symmetric* with respect to the n arguments, i.e., is not altered by any *interchange* between x, y, z, …; moreover the coefficients of the leading terms, x^p, y^p, z^p, etc., are equal to 1. Euler, and Leibnitz before him, expressed this effectively in the identity

$$(16) \qquad (x + y + z + \ldots w)^p - (x^p + y^p + z^p + \ldots + w^p) =$$
$$S\,(x, y, z, \ldots w)$$

The polynomial S is not only symmetric, but *homogeneous,* which means that the exponents of its *component monomials,* $M\, x^\alpha y^\beta z^\gamma \ldots w^\nu$, may range in value from 0 to p – 1, but are subject to the condition

$$(17) \qquad\qquad \alpha + \beta + \gamma + \ldots + \nu = p$$

The coefficients M of these monomials are positive integers, called *multinomial coefficients of order p.*

Now, these *multinomial integers* may be expressed in a very elegant form as "pseudo-fractions" in terms of factorials. It may be shown, indeed, either by *mathematical induction* or by *combinatorial arguments,* that if we denote the coefficient of $x^\alpha y^\beta z^\gamma \ldots w^\nu$ by the symbol $(\alpha, \beta, \gamma, \ldots \nu)$ then

$$(18) \qquad (\alpha, \beta, \ldots \nu) = \frac{(\alpha + \beta + \ldots + \nu)!}{\alpha!\beta! \ldots \nu!} = \frac{p!}{\alpha!\beta! \ldots \nu!}$$

provided that we *replace 0! where it occurs, by 1.*

I call the latter a *pseudo-fraction*. The artifice first appeared in print in Jacob Bernoulli's posthumous work *Ars Conjectandi*, published in 1713, a work which we would class today as a treatise on *combinatorial analysis* and *theory* of *probabilities*. It is, admittedly, a moot question whether it was he who had first conceived the formula, or his brother John, or Leibnitz, or any of their numerous correspondents.

How the multinomial theorem works in practice is illustrated on a *trinomial of order 5*. The result is

$$(19) \quad \begin{cases} (x + y + z)^5 = x^5 + y^5 + z^5 + 5(x^4y + xy^4 + y^4z + \\ yz^4 + z^4x + xz^4) + 10(x^3y^2 + x^2y^3 + y^3z^2 + \\ y^2z^3 + z^2x^3 + z^3x^2) + 20(x^3yz + y^3zx + z^3xy) + \\ 30(xy^2z^2 + yz^2x^2 + zx^2y^2) \end{cases}$$

Details are shown in the following table; the computations are checked by substituting $x = y = z = 1$ in the identity.

Type of Term	Number of Terms	Coefficient	Check
x^5	3	1	$1 \times 3 = 3$
x^4y	6	$5!/(4!) = 5$	$5 \times 6 = 30$
x^3y^2	6	$5!/(3!2!) = 10$	$10 \times 6 = 60$
x^3yz	3	$5!/(3!) = 20$	$20 \times 3 = 60$
x^2y^2z	3	$5!/(2!2!) = 30$	$30 \times 3 = 90$
	21		$3^5 = 243$

On Fermat's Little Theorem

The most simple and direct statement of the theorem is this: *If R is any positive integer, and p any prime, then $R^p - R$ is divisible by p.* Thus for $R = 2$:

$$(20) \quad 2^2 - 2 = 2, 2^3 - 2 = 3 \cdot 2, 2^5 - 2 = 5 \cdot 6, 2^7 - 2 = 7 \cdot 18$$

This special case of the *Little Theorem* was known to the ancient Chinese.

The theorem is often stated in a slightly different form; namely, *if the prime, p, does not divide R, then it divides $R^{p-1} - 1$.* Thus, for example, if p is a prime other than 2 or 5, then it divides $10^{p-1} - 1 = 999 \ldots 99$. This means that for a properly chosen number of digits the cryptogram 999 ... 9 is divisible by any prime other than 2 or 5, or, for that matter, by an integer Q which is *not divisible* by 2 or 5. This circumstance has an important bearing on the criteria of divisibility by Q, and also on the period of the decimal expansion of 1/Q.

The *converse of the little theorem is not generally true*, as the following "counter-examples" show:

I. $341 = 11 \cdot 31$ and is, consequently, *not a prime;* and yet 341 divides $N = 2^{340} - 1$, because one of the divisors of N is $2^{10} - 1 = 31 \cdot 33 = 3 \cdot 341$.

II. $121 = 11^2$ is *not a prime*, and yet $N = 3^{120} - 1$ is divisible by 121, because $3^5 - 1 = 242 = 2 \cdot 121$.

The history of the Little Theorem is most interesting. Fermat communicated it without proof to his friend Frenicle in a letter dated 1640; it appeared in print among the collected essays of Fermat, published posthumously by his son about 1660. Apparently, it made little impression on the mathematicians of the time. Thus when forty years later the theorem was rediscovered by Leibnitz, the latter could utter with impunity: "Here is something that no Analyst before me knew, a truly general formula for Prime Numbers." The phrasing implies a belief in the truth of the converse of the little theorem, which is puzzling indeed, for it is hard to believe that Leibnitz was not aware of the simple counter-examples given above.

Ironically, the Leibnitz contribution to Fermat's little theorem suffered the fate of the original. At any rate, in the decade between 1730 and 1740, Euler wrote much on the subject, but at no time did he mention Leibnitz, despite the fact that in one of his proofs Euler derived the theorem as a direct corollary to the *multinomial formula,* as did Leibnitz before him.

Assume that the order p of a multinomial is a *prime:* then, since the exponenets, $\alpha, \beta, \gamma \ldots$, are less than p, the denominator of the pseudo-fraction contains no factor which could cancel out p, and the integer $(\alpha, \beta, \gamma \ldots)$ is in this case divisible by p. Hence this important lemma: *all multinomial coefficients of prime order p are multiples of p.* In virtue of this theorem, the multinomial identity of the preceding article takes on the form

(21) $(x + y + z + \ldots + w)^p - (x^p + y^p + z^p + \ldots + w^p) =$
$$pH(x, y, z, \ldots w)$$

where $H(x, y, z, \ldots w)$ is a polynomial with positive integral coefficients.

The number-theoretical implications of this lemma are extensive, and the Little Theorem is one of the most simple and important of these. Set in identity (21) $x = y = z = \ldots = w = 1$, $H(1, 1, 1, \ldots 1) = N$, and it becomes

(22) $\qquad n^p - n = Np$ or $n(n^{p-1} - 1) = Np$

Thus, if the prime p does not divide n, it divides $n^{p-1} - 1$.

This, substantially, is Leibnitz's proof of Fermat's theorem. One of Euler's proofs differs from it in phraseology only; the other is an excellent illustration of mathematical induction. The crux of the proof is to view *p as fixed and R as variable.* For $R = 1$ we have $1^p - 1 = 0$, which establishes the *induction step.* Assume now that the theorem holds for some value of R, i.e., that $R^p - R = Ap$ where A is some positive integer, and consider the expression $(R + 1)^p - (R + 1)$. In virtue of the lemma mentioned above $(R + 1)^p - R^p - 1 = Bp$, provided p is a prime. Consequently

$\qquad (R + 1)^p - (R + 1) = R^p - R + Bp = (A + B)p.$ Q.E.D.

On Wilson's Theorem

The story of this proposition illustrates the role of paraphrase in mathematical reasoning. In 1770 Edward Waring published a book under the title *Meditationes Algebraicae*. One passage from this work reads: "If p is a prime, then the quantity

$$(23) \qquad \frac{1 \cdot 2 \cdot 3 \ldots \cdot (p-1) + 1}{p}$$

is a whole number. ... This elegant property of prime numbers is the discovery of the eminent John Wilson, a man well versed in matters mathematical." This glowing tribute to Wilson should not be taken too seriously, for there is evidence that this was Waring's way of paying off a "political" debt. Waring adds: "Theorems of this kind will be very hard to prove, because of the absence of a notation to express prime numbers." Commenting on this passage Gauss uttered his celebrated bon mot on "*notationes versus notiones*," meaning that in questions of this kind it was not *nomenclature*, but *conception* that mattered.

Despite Waring's pessimistic forecast, the theorem was proved independently by Euler and Lagrange within a few years of its announcement. The methods used by these masters transcend by far the scope of the problem which has inspired them, and for this very reason were indirect and involved. By contrast, the Gaussian proof given below is so simple and direct that even a reader with a modicum of mathematical training should understand and appreciate it.

Wilson's theorem is equivalent to the relation

$$(24) \qquad (p-1)! = Wp - 1$$

where W is an integer, and, since we shall confine our study to values of p greater than 3, we can take W greater than 1. Let us accordingly set $W = G + 1$ which turns (24) into

$$(24') \qquad (p-1)! = Gp + (p-1)$$

The theorem can now be paraphrased as follows: *if p is a prime, the remainder of division of (p − 1)! by p is (p − 1).* On the other hand, $(p − 1)! = (p − 1)(p − 2)!$, and so the problem is reduced to proving that *the remainder of division of (p − 2)! by the prime, p, is 1.*

To prove this last statement, Gauss uses a device which for want of a better name I shall call *pairing.* I present the method in a table using the case $p = 11$ as an example. It is, for all intents and purposes, a *10 by 10 multiplication table,* except that the entries are not the products, but their *residues modulo II.*

1	1	2	3	4	5	6	7	8	9	10
2	2	4	6	8	10	1	3	5	7	9
3	3	6	9	1	4	7	10	2	5	8
.	.	.	.	.	.	.	.	.	.	.
.	.	.	.	.	.	.	.	.	.	.
9	9	7	5	3	1	10	8	6	4	2
10	10	9	8	7	6	5	4	3	2	1

The array has the following properties: First, *to any two equal entries corresponds two integers the difference of which is divisible by p.* Second, *no two entries in the same row or column can be equal.* Third, since every row has $p − 1$ entries, *every one of the p − 1 residues is represented in each row, and only once.* Fourth, in particular, *the residue 1 is present in each row;* which means that we can associate with each integer k of the sequence 2, 3, 4 … $p − 2$, another integer K such that the residue of $k \cdot K$ is equal to 1.

In the case $p = 11$ we can regroup the product as follows

$$2 \cdot 3 \cdot 4 \cdot \ldots \cdot 8 \cdot 9 = (2 \cdot 6)(3 \cdot 4)(5 \cdot 9)(7 \cdot 8)$$

The product of each pair when divided by 11 leaves a remainder 1. The same, therefore, holds for the product of all these pairs. Hence, res(9!) = 1 and res(10!) = 10, i.e., 10! + 1 = 11 N, which is Wilson's theorem for $p = 11$.

Adapting this argument to the general case, we reach the conclusion: *if p is a prime, then the p − 3 factors, 2, 3, 4, ..., (p − 2) may be arranged into* $\frac{1}{2}(p − 3)$ *associate pairs;* the residue of each pair modulo p is 1; the residue of the products of all *associate pairs* is, therefore, also 1; from this we infer that when p is a prime, the residue (mod p) of (p − 1)! is p − 1, and this is Wilson's theorem in its Gaussian version.

The table on page 53 is built on a different paraphrase of the Wilson theorem. Let w_p be the remainder of division of $[(p − 1)! + 1]$ by p: I shall call w_p *the Wilson index of the integer p.* Wilson's theorem states that the *index of a prime is 0;* conversely, *if the index of p is 0, then p is prime.* What happens when p is composite? The answer is: *The Wilson index of any composite integer other than 4 is equal to +1.*

The last statement is but a paraphrase of the following theorem: *If p is a composite integer greater than 4, then p divides (p − 1)!.* Proof: Assume first p composite, but not the square of a prime; then it is the product of two distinct integers, both less than p − 1, and therefore divisors of (p − 1)!. Assume next $p = q^2$, where q is an odd prime: then q is less than p − 1, and so is 2q. Hence both q and 2q enter into (p − 1)!; it follows that the latter is divisible by $2q^2$, and, consequently, by p.

If we exclude integers less than 4, we can sum up the preceding discussion in the words: *the index of a prime is 0, the index of a composite integer is 1.* Accordingly, we can represent the sequence of natural numbers as an infinite binary fraction in which the digit 1 designates a composite integer and the digit 0 a prime.

$$(25) \begin{cases} .010111 \quad 0101110101 \quad 1101111101 \quad 0111110111 \quad 0101110111 \\ 1101111101 \quad 0111110111 \quad 0101111101 \quad 1101111101 \quad 1111110111 \end{cases}$$

The first digit of this fraction represents p = 5, the last, p = 100.

On a Problem of Lagrange

The *Null Factor Law,* of which I shall have much to say in the next group of topics, stipulates that *a product cannot be equal to zero unless at least one of its factors is zero;* or in symbols, the relation uv = 0 implies either u = 0, or v = 0, or u = v = 0. The number theoretical counterpart to this law is: *If a prime p divides a product of integers, then it divides at least one of the factors.* By means of a terminology and symbolism introduced by Gauss, the resemblance between the two properties can be made even more striking.

If the integers a and b have the same remainder of division by p, then Gauss calls a and b *congruent modulo p* and writes $a \equiv b$ (mod p). In particular, $c \equiv 0$ (mod p) means that c is *divisible* by p. With this notation we can paraphrase the analogy to the Null Factor Law into the following form: If p is a prime, and $uv \equiv 0$ (mod p), then either $u \equiv 0$ (mod p), or $v \equiv 0$ (mod p), or $u \equiv v \equiv 0$ (mod p).

Consider now a polynomial $G(x)$ of degree n. For what integral values of x, if any, is $G(x)$ divisible by a given prime p? This is Lagrange's problem, the Gaussian paraphrase of which reads: *Determine the solutions of the congruence $G(x) \equiv 0$ (mod p).* At this junction the analogy between equation and congruences hits a snag: If *a* is a solution of any congruence modulo *p*, then $a + np$ is also a solution, whatever n. Lagrange met this difficulty by designating the *least positive* term of the progression $a + np$ as representative of the whole progression. Gauss calls these *minimal solutions roots.* For example, the congruence $x^2 + 1 \equiv 0$ (mod 5) has an infinite number of solutions: 2, 7, 12, 17, ...; 3, 8, 13, 18, ... but only *two distinct roots,* 2 and 3.

With this terminology, Lagrange's two fundamental theorems read: First, *A congruence of degree n can have at most n distinct roots;* and second, *If a polynomial of degree n is divisible by p for more than n non-congruent values of x, then it is divisible by p for any value of x.*

This last proposition led Lagrange to the discovery of a remarkable relation between Fermat's Little Theorem and that of Wilson. Consider the polynomial

(26) $$G(x) = (x - 1)(x - 2)(x - 3)$$
$$\dots [x - (p - 1)] - (x^{p-1} - 1)$$

where p is a prime. We find

$$G(1) = 0, G(2) = 2^{p-1} - 1, G(3) = 3^{p-1} - 1, \dots$$
$$G(p - 1) = (p - 1)^{p-1} - 1$$

In virtue of Fermat's Little Theorem, all these values are *divisible by the prime p;* thus the congruence $G(x) \equiv 0 \pmod{p}$ admits $p - 1$ *non-congruent* solutions. On the other hand, the degree of the polynomial $G(x)$ is only $p - 2$. It follows that $G(x)$ is divisible by p for all values of x, and, in particular, for x = 0. Hence

(27) $$G(0) \equiv [(p - 1)! + 1] \equiv 0 \pmod{p}$$

which is Wilson's theorem.

The Eratosthenes sieve shown on page 48, or the binary fraction, (25), exhibits the erratic distribution of primes among the first one hundred integers. The irregularity persists and becomes even more disconcerting when we penetrate deeper into the natural sequence of numbers. Those who would seek rhyme or reason in this bizarre array should weigh some figures and facts and fancies that failed.

There are 5 primes between 101 and 113, but none between 114 and 126. We find 23 primes between 1 and 100, and 21 primes between 101 and 200; but between 8,401 and 8,500 there are only 8 primes, and these 8 are crowded in the interval 8,418 to 8,460. And lest the reader get an erroneous notion, let me hasten to add that there are 13 primes between 89,501 and 89,600.

Reliable factor tales exist for the first 10,000,000 integers: there are 664,580 primes among these. Beyond this, one must depend on appraisal theories and formulae. The first such formula was given by Legendre in 1808. Basing his deductions on an empirical study of Euler's factor table, he arrived at the approximation formula

(28) $x / \pi(x) \cong \ln(x) - B$

where $\pi(x)$ designates *the number of primes which are less than or equal to x*; lnx is the natural logarithm of x, and B is a slowly varying magnitude, averaging about 1.08.

An alternative approximation formula was proposed by Gauss:

(29) $\pi(x) \cong \int_{2}^{x} dx / \ln(x)$

The integral has since received the name *Logarithmic Integral,* and is tabulated as Li(x).

Gauss conjectured that the *ratio $\pi(x)/Li(x)$ tends to 1 when x tends to infinity,* and this conjecture was proved by the Belgian mathematician De la Valleé-Poussin in 1896. The Legendre formula was turned by Chebyshëv from an empirical law into a mathematical conjecture. In 1848 this Russian mathematician proved that if the ration of $\pi(x)$ to (x/lnx) did approach a limit, then this limit must be 1. The existence of the limit was proved by the French mathematician Hadamard in 1896. The Chebyshëv-Hadamard proposition is now known as the *Prime Number Theorem.*

Hadamard proved the Prime Number Theorem by methods of *Analytical Number Theory*, i.e., by means of infinite processes. Many experts maintained that a direct arithmetic proof of the proposition would never be forthcoming, but in 1950 such a proof was given through the combined efforts of Selberg and Erdös.

How careful one must be in drawing inferences from the Prime Number Theorem will be gleaned from the following example: The sequence of one million consecutive integers,

(30) 1,000,001! + 2, 1,000,001! + 3, ... 1,000,001! + 1,000,001,

constitutes a "solid block" of composite numbers, because N! + k is divisible by k, as long as k is greater than 1 and less than N + 1. It follows that the number of primes in that interval is zero, while the approximation formulae would lead one to believe that primes exist in any definite interval sufficiently large.

Formulae for Primes

The quest dates back to Fermat. Its specific aim has been to determine a function $G(x)$ which would yield primes for *all* integral values of the argument x; the function sought is to involve only additions and multiplications so that any step in the process, initial, intermediate, or final, would bear on *integers exclusively*. It would be convenient to call such functions *arithmetic* or *integral;* but as much as these terms have already been preempted, I propose to designate them as *generic*. The prototype of generic functions is a *polynomial with integral coefficients*. Other examples are

(31) $a^x + b, a^{G(x)} + H(x), G(x)^{H(x)} + K(x)$

where a, b, are integers, and G, H, K, are polynomials with *positive* integral coefficients.

In spite of the restrictions imposed, the variety of generic functions is immense, and so it is natural to inquire whether at least one of these functions would represent primes for all integral values of x. Well, thus far, no such function has been discovered; but neither has the possibility of the existence of such a formula been disproved. On the other hand, it has been proved that certain types of generic functions cannot represent primes exclusively. One of the earliest propositions of this kind is a theorem of Euler to the effect that *any generic polynomial must take on composite values for at least one value of the argument.*

Euler's theorem rests on the following algebraic lemma: If $P(x)$ is any polynomial, then the polynomial

$$Q(x) = P[x + P(x)]$$

admits $P(x)$ as a factor. Thus, set $P(x) = x^2 + 1$, then

(32) $Q(x) = [x + (x^2 + 1)]^2 + 1 = (x^2 + 1)^2 + 2x(x^2 + 1) +$
    $x^2 + 1 = (x^2 + 1)(x^2 + 2x + 2)$

The proof of the general lemma follows the line of this example and is quite formal; so I leave it to the reader.

Suppose next that $P(x)$ is a generic polynomial; let *a* be any integer and set $b = P(a)$; then, according to the lemma, the integer $P(a + b)$ is divisible by b and is therefore composite, provided $b \neq 1$. Observe, that if n be the degree of the polynomial $P(x)$, then according to the fundamental theorem of algebra there exist at most n values for which $P(x) = 1$, consequently the Euler theorem can be amplified to read that *any generic polynomial assumes an infinite number of composite values.*

Closely allied to the preceding question is another: Do generic functions exist which, without representing primes *exclusively*, may take on an *infinite number* of prime values?

Euclid's theorem suggests an affirmative answer to this question; for, obviously, it may be paraphrased to read: *The linear function* $G(x) = 2x + 1$ *may assume an infinite number of prime values.* Euler and Legendre took it for granted that the same holds for such arithmetic progressions as $3x + 1$, $3x + 2$, $4x + 1$, $4x + 3$, and Gauss conjectured that *any arithmetic progression contains an infinity of primes* provided that the *first term and the difference of the progression are relatively prime,* or, which amounts to the same thing, that the linear function

(33) $$G(x) = px + q$$

will take on an infinite number of prime values provided the integers p and q are relatively prime. The proposition was proved in 1826 by Lejeune Dirichlet, who used in the process very subtle analytical arguments.

On the other hand, all attempts to generalize Dirichlet's methods to *non-linear* functions have thus far been unsuccessful. As a matter of fact, despite the prodigious strides which have been made in this field since the days of Dirichlet, not a single non-linear generic function is known of which it may be asserted with mathematical certainty that it can take on an infinity of prime values.

Here are some classical examples which bring out the present status of the problem:

I. *The quadratic* $G(x) = x^2 + 1$. A necessary condition that $G(x)$ be prime is that x ends in the digits 4 or 6. This gives rise to the values 17, 37, 197, 257, ... The first 66 terms of this sequence contain 12 primes. Many larger primes of the type are known, but whether the aggregate is *finite* or *infinite* is still a moot question.

II. *The Mersenne function* $G(x) = 2^x - 1$. As was mentioned before, the prime values of $G(x)$ are known as Mersenne numbers, and only 17 such integers are known today. The conjecture that there is an infinite number of Mersenne primes remains unproved.

III. *The Fermat function* $G(x) = 2^x + 1$. Since any integer is of the form $x - 2^P M$, where M is odd, a *necessary* condition for the *primality* of $G(x)$ is $M = 1$. This leads to the Fermat numbers mentioned on page 50. It may be added here that these integers play an important role in geometrical construction because of a fundamental theorem of Gauss to the effect that a *regular polygon of n sides can be erected by straightedge and compass if, and only if, n is a Fermat prime or a square-free product of Fermat primes.* Here again the question as to whether the aggregate of Fermat primes is finite or infinite is one of the unsolved problems of number theory.

Pythagorean Triples

These integers have, in modern times, led to many number-theoretical discoveries, and also to many perplexing problems some of which still await solutions. The impetus to this modern development was provided by Fermat, who in his marginal notes stated without proof many theorems involving these integers, theorems which a century or so later were confirmed and amplified by Euler, Lagrange, Gauss, and Liouville.

To facilitate exposition I shall call any integral solution of the equation

$$(34) \qquad\qquad x^2 + y^2 = R^2$$

a *triple:* x and y are the *sides* of the triple, R its *hypotenuse.* The triple is *primitive,* if the elements have no divisors in common, *imprimitive* otherwise. If (x, y, R) is a triple, then obviously (nx, ny, nR) is also one; thus, to any primitive triple may be assigned an infinity of imprimitive triples, and this puts primitive triples in the foreground. Certain properties of these latter flow directly from the definition, and were known to the ancients. The most important of these are: *(a) the aggregate of primitive triples is infinite, (b) any two elements of a primitive triple are relatively prime, (c)* the sides of a primitive triple are of *opposite parity,* while the *hypotenuse is always odd.*

The point of departure in Fermat's approach is a proposition implicitly used by Diophantus, Fibonacci, and Vieta; namely, that the *hypotenuse of a Pythagorean triple may be represented as a sum of two squares.* In symbols, if R is an "admissible" hypotenuse, then integers p and q exist such that

(35) $$p^2 + q^2 = R$$

That this condition is *sufficient* follows from the identity

(36) $$(p^2 + q^2)^2 = (p^2 - q^2)^2 + (2pq)^2$$

Thus equation (34) is satisfied by the triple

(37) $$x = p^2 - q^2, y = 2pq, R = p^2 + q^2,$$

where p and q are any integers whatsoever. The proof that *this sufficient condition is also necessary* hinges on a rather subtle argument which demands more space than I can spare here.

Equations (37) hold for imprimitive triples as well as primitive; however, it is easy to see that by restricting the *parameters* p and q to be *relatively prime* and of *opposite parity,* one automatically eliminates all imprimitive triples. This is clearly exhibited in the accompanying table, where every entry is obtained by adding an even to an odd square. The canceled entries lead to *imprimitive triples,* inasmuch as the generating parameters are not co-prime; the underlined elements, on the other hand, represent *prime hypotenuses* which, as we shall presently see, play a fundamental role in the Fermat approach to the Pythagorean problem.

	4	16	36	64	100	144	196	...
1	5	17	37	65	101	145	197	...
9	13	25	45	73	109	153	205	...
25	29	41	61	89	125	169	221	...
49	53	65	85	113	149	193	245	...
81	85	97	117	145	181	225	277	...
121	125	137	157	185	221	265	317	...
169	173	185	205	233	269	313	365	...
...	...	...	...	...	...	...	...	...

Now, to recognize that a given odd number can be represented as the sum of two squares is a thorny problem. To be sure, the sum of an odd and even square is always of the form $4n + 1$, which automatically "disqualifies" any term of the progression,

$$4n - 1: \qquad 3, 7, 11, 15, 19, 23, \ldots$$

Unfortunately, however, merely belonging to the progression

$$4n + 1: \qquad 5, 9, 13, 17, 21, 25, 29, 33, 41, \ldots$$

is not sufficient. Thus for example, 9 cannot be resolved into the sum of two squares, neither can 21 or 33. Indeed, it was this difficulty of finding a sufficient condition that had blocked the efforts of Fermat's brilliant predecessor, Vieta.

Fermat overcame this difficulty by distinguishing between *prime* and *composite* hypotenuses. In what follows I shall refer to the corresponding triples as *primary* and *compound*. Examples of primary triples are (3, 4, 5), (5, 12, 13), and (15, 8, 17); examples of compound are (7, 24, 25), (63, 16, 65), and (33, 56, 65). In his proof Fermat invoked one of his own celebrated marginal theorems; namely, that *any prime of the form $4n + 1$ can be represented as a sum of two squares, and, what is more, the representation is unique.* As usual, the marginal note was not accompanied by a proof; however, in a letter to Roberval, Fermat stated that he possessed a rigorous demonstration, based on his method of *indefinite descent*. Euler, 125 years later, furnished the first published proof of the lemma using the principle of descent as basis.

The immediate consequence of the theorem is that *any prime of the form 4n + 1 is an admissible hypotenuse*. How about composite integers of this type? This question is answered in Fermat's marginal notes as follows:

Let R = $p^\alpha q^\beta r^\gamma$ ... where p, q, r, ... are *odd primes* and $\alpha, \beta, \gamma,$ ... positive integers. Then we must distinguish four cases:

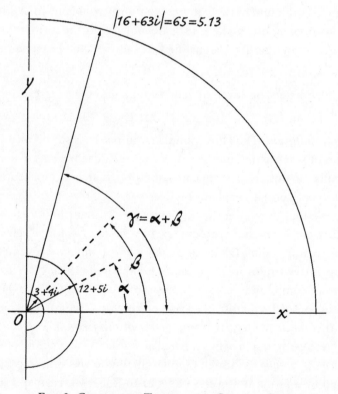

FIG. 3. COMPOUND TRIPLES AND COMPLEX INTEGERS

I. *All the prime divisors are of type 4n + 1.* The Pythagorean equation then admits at least one *primitive* solution.

II. *All the prime divisors are of type 4n − 1.* Equation (34) admits *no solution.*

III. *At least one of the prime divisors is of type 4n + 1, while any prime divisor of type 4n − 1 enters in the product to an even power.* Equation (34) admits *imprimitive solutions* only.

IV. *Any one of the prime divisors of type 4n − 1 enters to an odd power.* The equation admits *no solution,* primitive or otherwise.

Fermat's approach shows that not only does the Pythagorean equation admit of an infinity of primitive solutions, as was surmised by the ancients, but of an infinity of *primary solutions.* These may be used as building stones to derive all other solutions, primitive and imprimitive. As to the compounding process, it is purely formal, being fully equivalent to *multiplication of complex numbers* (see Fig. 3). We may, indeed, interpret the "complex integer" $x + iy$ as the sides of a triple and its absolute value $R = |x + iy|$ as its hypotenuse. The product of two such complex integers is

(38) $$(x + iy)(x' + iy') = (xx' - yy') + i(xy' + x'y)$$

and the elements of this new integer give rise to a new solution of equation (34) in virtue of the identity

(39) $$(xx' - yy')^2 + (xy' + x'y)^2 = (x^2 + y^2)(x'^2 + y'^2)$$

What is the present-day status of the problem? Suppose you were asked to determine the triples which admit some given integer R for hypotenuse. The last two digits reveal that R is of the type $4n + 1$, and so your next step is to calculate the prime divisors of R, a formidable task if R is large enough. Assume, then, for the sake of argument, that you have succeeded to prove that R is a prime. Then there exists a unique triple of which R is the hypotenuse, and the quest is reduced to solving the *Diophantine equation:* $p^2 + q^2 = R$. This latter problem was, in turn, reduced by Lagrange to the study of a *continued fraction* associated with $\sqrt{R}$. Other methods have since been invented to cope with the problem; generally speaking, however, the problem of determining the *component squares* of an integer is at least as difficult as resolving an integer into prime divisors. Under the circumstances, it would hardly be fair to claim that the ancient problem has been exhaustively solved.

An Euler Episode

In a paper published in 1774, Euler listed several large primes, among which was the integer 1,000,009. In a subsequent paper Euler admitted his error and gave the prime divisors of that integer as

(40) $1,000,009 = 293 \times 3,413$

He remarked that at the time the first paper appeared he had been under the impression that 1,000,009 admitted of a *unique representation in squares,* namely, $1,000,009 = 1,000^2 + 3^2$; but that he had discovered since a *second partition,* $235^2 + 972^2$, which revealed the composite character of the number.

Euler then proceeded to calculate the divisors of 1,000,009 by a method patterned along the proof of a theorem which he presented in an earlier paper. The proposition, which harks back to a marginal note of Fermat, states that *if an odd integer, R, is susceptible of more than one partition into two squares, then R is composite.* Euler's method rested on a simple lemma, so simple, indeed, that most readers would be inclined to accept it as self-evident. If A/B and a/b are two fractions, the latter *in lowest terms,* then *the equality $A/B = a/b$ implies the existence of an integer n, such that $A = na$ and $B = nb$.*

Let us now restate the problem. Given that

(41) $$R = p^2 + q^2 = p'^2 + q'^2,$$

show that R is composite and determine its divisors. First we rewrite (41) as a proportion

$$p^2 - p'^2 = q'^2 - q^2 \quad \text{or} \quad \frac{p + p'}{q' - q} = \frac{q' + q}{p - p'} = \frac{a}{b}$$

where a and b are assumed relatively prime. Invoking next the above lemma, we obtain the four relations

$$p + p' = ma \qquad\qquad q' + q = na$$
$$q' - q = mb \qquad\qquad p - p' = nb$$

By squaring and adding these we obtain:

(42) $$(m^2 + n^2)(a^2 + b^2) = 2(p^2 + q^2 + p'^2 + q'^2) = 4R$$

From this we conclude that $a^2 + b^2$ is one of the sought divisors of R if a and b are of opposite parity; if, on the other hand, both a and b are odd, then $\frac{1}{2}(a^2 + b^2)$ is a divisor of R.

When this method is applied to $R = 1,000,009 = 1000^2 + 3^2 = 235^2 + 972^2$, we are led to the proportion

$$\frac{1972}{238} = \frac{232}{28} = \frac{58}{7}$$

Hence $a = 58, b = 7, a^2 + b^2 = 3413$ which is one of the two divisors sought.

How Euler stumbled on the second partition is not on record. The prodigious number sense and the phenomenal memory of this master calculator are astounding enough, but what makes the mystery even more profound is that at the time the incident occurred Euler was in his seventies, and that he had been totally blind for more than a decade, and partially blind long before.

On Perfect Numbers

The following is a free version of Euler's proof of Euclid's theorem that *any even perfect number is of the form $2^{p-1}(2^p - 1)$,* where the *odd factor is prime;* and that, conversely, if M is a *Mersenne number,* i.e., a prime of the form $2^p - 1$, *then $2^{p-1}M$ is a perfect number.*

Any even integer may be put in the form $P = 2^{p-1}M$ where p is greater than 1 and M odd. Denote by S the sum of all odd divisors of P, by Σ the sum of all divisors including P itself, and set $\Sigma = S - M$. Observe that Σ is 1 if M is prime, and is greater than 1 if M is composite.

The two sums Σ and S are connected by the general relation

(43) $\Sigma = S + 2S + 2^2S + \ldots + 2^{p-1}S = S(2^p - 1).$

Assuming next that P is a perfect number, we have, in addition, $\Sigma = 2P = 2^pM$. Identifying these two values of Σ, we are led to the equations

(44) $S = 2^p\Sigma$ $M = (2^p-1)\Sigma$

The first shows that S is even, from which we conclude that Σ is odd. The second relation presents us with an alternative. Either Σ is 1, and M is a prime; or M is composite and Σ a divisor of M. The second hypothesis would mean that P has at least three odd divisors, 1, M, and Σ, which is impossible since the sum of these three is $1 + S$, while the sum of *all odd divisors* is S. It follows that Σ is 1, and M is prime; equation (44) then shows that $M = 2^p - 1$.

In the same order of ideas is Sylvester's proof that *an odd perfect—if one existed—must have at least three prime divisors.* Let us first dispose of the possibility of a single prime divisor. If x be a prime, and P a perfect integer of the form x^m, then

(45) $x^m = 1 + x + x^2 + \ldots + x^{m-1} = \dfrac{x^m - 1}{x - 1};$

this leads to the relation $x^m(2 - x) = 1$, which is patently impossible.

Nor can a perfect integer be of the form $P = x^m y^n$ where x and y are *distinct odd primes*. Indeed, let X and Y be the sums of the divisors of x^m and y^n respectively and let Σ be the sum of all the divisors of P. Then, since P is assumed perfect, $\Sigma = 2P$. But, on the other hand, $\Sigma = XY$; equating these two values of Σ we obtain

$$(1 + x + x^2 + \ldots + x^m)(1 + y + y^2 + \ldots + y^n) = 2x^m y^n$$

or

$$(x^{m+1} - 1)(y^{n+1} - 1) = 2(x - 1)(y - 1)x^m y^n.$$

To prove that this relation is impossible, Sylvester puts it in the forms

$$(46) \qquad \left(1 - \frac{1}{x^{m+1}}\right)\left(1 - \frac{1}{y^{n+1}}\right) = 2\left(1 - \frac{1}{x}\right)\left(1 - \frac{1}{y}\right)$$

and observes that the right member attains its least value for $x = 3$, $y = 5$, and therefore never falls below 16/15, while the left member never exceeds 1.

On Roots and Radicals

"Any cubic or biquadratic problem can, in the last analysis, be reduced either to the trisection of an angle or to the doubling of a cube."

—Vieta

The Phoenician Bequest

The generally accepted theory that both the Greeks and the Jews owe their systems of writing to the Phoenicians is strongly supported by the similarity in the names of the symbols: compare the Greek *alpha, beta, gamma,* with the Hebrew *aleph, beth, gimmel.* Significant also is the circumstance that by adapting the Phoenician method to their own needs, the Jews as well as the Greeks had committed themselves to the *dual* character of the system: indeed, every letter of the alphabet was not only the record of a sound, but the symbol of a number. One effect of this was the *Gematria* discussed in Chapter 3; however, the inherent duality of the Greek script had other consequences which influenced the subsequent course of mathematics in more than one way.

Now, whether *phonetic writing* was the invention of some anonymous Phoenician genius, or whether the Phoenicians had inherited their script from an earlier civilization, one thing is certain: the principle was so vast an improvement over all previous methods of recording experience that it underwent no significant change after adoption by the Greeks. One great advantage is that the script requires so few symbols that even a man of average mentality can readily memorize the letters in a prescribed order. The *ordinal* aspect of the alphabet makes it a natural counterpart of the counting process. But the correlation is by no means complete, since the Greek alphabet contains twenty-two letters, while positional numeration and the decimal structure of the spoken number language require but ten symbols. The Greeks used every one of the letters of their alphabet as numerals, and this *embarras de richesses,* as the French would put it, was undoubtedly a serious stumbling block to the discovery of the principle of position, without which no real progress in arithmetic was possible.

This dual character of the Greek alphabet had also a retarding effect on the development of other branches of mathematics. We know today that no substantial advance in algebra was possible until means were invented to designate *abstract* magnitudes, known or unknown, variable or constant, and in particular the indeterminate constants which we call today *coefficients* and *parameters.* Before this could be achieved, the letter had to be freed from its numerical value. The introduction of Arabic numerals was an important step in that direction. Still, so great is the power of tradition, that little progress was made in the nearly four hundred years which separated Fibonacci from Vieta. To be sure, Italian mathematicians of the sixteenth century did devise methods for solving cubic and quartic equations; however, their solutions were not expressed in general terms but exhibited on typical numerical examples. Indeed, it was not until the publication in 1592 of Vieta's work on *literal notation* that the letter was finally emancipated from the shackles which the Phoenicians had imposed upon it.

The Harriot Principle

This article deals with the role of zero in that branch of algebra which is known as Theory of Equations. To be specific, we are concerned here with a procedure which consists in *transposing all terms* of an equation to one side of the equality sign and writing it in the form $P(x) = 0$, where $P(x)$ is a *polynomial*.

I call this procedure *Harriot's Principle*. Thomas Harriot, geographer, one-time tutor to Sir Walter Raleigh, and first surveyor of the Virginia territory, had no mathematical reputation during his lifetime. As a matter of fact, the principle in question did not see the light of day until 1631, when Harriot had been dead for nearly ten years. Even then the credit for this innovation was bestowed upon another man, for soon after the publication of Harriot's *Praxis,* Descartes' book on analytic geometry appeared, in which the philosopher made free use of Harriot's ideas, but, true to form, never mentioned the source. So great was the renown of Descartes that for nearly a century the epoch-making principle was generally regarded as a Cartesian achievement.

I use the word *epoch-making* advisedly, for in spite of its utmost simplicity, the principle ranks in historical importance with Vieta's introduction of literal notation. In the first place, by *reducing the solution of an equation to the factoring of a polynomial,* the Harriot approach brought about a vast improvement in equational technique. In the second place, it led to the study of the relations which connect the roots of an equation to its coefficients and, for this reason, foreshadowed many subsequent theoretical developments, such as *symmetric functions, the fundamental theorem of algebra, and functions of a complex variable.*

To understand the source of this prodigious fertility, we must remember that the real and complex magnitudes to which the principle applied obey not only the formal laws of algebra, but are also subject to the *Null Factor Law*. Expressed in words, the latter states that *a product cannot be equal to zero unless at least one of its factors is zero*, or in symbols, the relation $uv = 0$ implies either $u = 0$, or $v = 0$, or $u = v = 0$.

But if it was this law that bestowed upon the Harriot principle its great power, why did neither Harriot nor the legion of mathematicians who applied his principle in the two centuries which followed him ever mention the Null Factor Law, even by indirection? The answer is that before one can make any statement, one must be able to conceive an alternative to the Null Factor Law; but in Harriot's day, when even the complex number was eyed with undisguised suspicion, this was out of the question. And I dare say that the very idea of setting up an algebraic code and then seeking out mathematical beings who would obey the code, would have been regarded by most mathematicians of the seventeenth and eighteenth centuries as the raving of a maniac.

Equation and Identity

The Harriot approach discussed in the preceding article brings out the remarkable kinship between *algebraic equations* and *polynomial functions*, and suggests that the two concepts are interchangeable. However, we shall presently see that the analogy is far from complete, and that it leads, if carried too far, to results which border on the absurd.

No difficulty arises if the given equation is defined unambiguously, i.e., when all the coefficients of the corresponding polynomial can be expressed in terms of the data of the problem. This, however, is rarely the case: many problems in mathematics, pure and applied, depend on *variable parameters*, and the coefficients of the equation to which any one of these problems may lead are not constants, but functions of these parameters. The result is that one is confronted not with a single equation, but with an *aggregate of equations.*

Thus, the *general quadratic function* in one variable is given by the formula

(47) $Q(x) = ax^2 + bx + c$

Assume that the indeterminate coefficients a, b, c, range over all rational values, positive, negative, or zero. Then the aggregate (Q) would include not only *bona fide* quadratic functions but also such functions as

(48) $\begin{cases} Q(x) = 0 \cdot x^2 + bx + c & (b \neq 0) \\ Q(x) = 0 \cdot x^2 + 0 \cdot x + c & (c \neq 0) \\ Q(x) = 0 \cdot x^2 + 0 \cdot x + 0 \end{cases}$

We interpret this graphically by saying that the function $y = Q(x)$ represents not only *parabolas,* but also *straight lines,* including lines parallel to the x axis, and the x axis itself.

When, however, we try to discuss the same special cases as equations, we find that we cannot indulge in such sweeping interpretations. Indeed, from the standpoint of pure algebra, the relation $0 \cdot x^2 + bx + c = 0$ *is not a quadratic,* but a *linear equation;* the relation $0 \cdot x^2 + 0 \cdot x + c = 0$ ($c \neq 0$) is not an equation but an *incongruity,* and the relation $0 \cdot x^2 + 0 \cdot x + 0 = 0$ is not an equation but a *tautology.* Moreover, since these difficulties are inherent in the *definition of the number zero,* they cannot be evaded by such formal devices as the Harriot procedure. We saw, indeed, that the latter would be useless without the Null Factor Law, which, in turn, is but a corollary to the conditions under which zero has been admitted to the number domain.

Thus a polynomial relation is not necessarily a *bona fide* equation: it may be an *incongruity* or an *identity.* Strangely enough, this very ambiguity can be turned into an effective means for proving identities. The method derives from the proposition that *if a polynomial relation of degree n in x is satisfied by more than n distinct values of x, then the relation is an identity,* i.e., is satisfied for *any value* of x.

Consider, for instance, the *quadratic*

$$P(x) = (b - c)(x - a)^2 + (c - a)(x - b)^2$$
$$+ (a - b)(x - c)^2;$$

we find by direct substitution that

$$P(a) = P(b) = P(c) = (a - b)(b - c)(c - a).$$

Thus the relation is satisfied by more than two values of x, and is, therefore, an *identity*, i.e.,

(49) $(b - c)(x - a)^2 + (c - a)(x - b)^2 + (a - b)(x - c)^2$
$$\equiv (a - b)(b - c)(c - a)$$

As a second example, consider the relation

(50) $$\frac{(x-a)(x-b)(x-c)}{(d-a)(d-b)(d-c)} + \frac{(x-b)(x-c)(x-d)}{(a-b)(a-c)(a-d)} +$$
$$\frac{(x-c)(x-d)(x-a)}{(b-c)(b-d)(b-a)} + \frac{(x-d)(x-a)(x-b)}{(c-d)(c-a)(c-b)} = 1$$

Denote the left member by $P(x)$ and observe that this cubic assumes the value 1 for *four* values of x. We find, indeed,

$$P(a) = P(b) = P(c) = P(d) = 1$$

and conclude that the relation is an identity.

On Cubics and Quartics

Lagrange's suspicion that the roots of equations of degree higher than the fourth *cannot* be generally expressed by means of radicals was confirmed by Abel and Galois. The demonstration lies beyond the scope of the present work; however, much insight into the nature of the problem may be gained by studying the methods used by Euler and Lagrange for cubic and quartic equations, methods which, incidentally, bring out the importance of the Harriot principle to equational technique.

If the identity

(51)
$$H(x) = x^3 + a^3 + b^3 - 3abx \equiv$$
$$(x + a + b)(x^2 + a^2 + b^2 - ax - bx - ab)$$

we view a and b as given and x as unknown, we can interpret the identity to mean that the *defective* cubic equation $H(x) = 0$ admits $x = -a - b$ for a root. Now, the general defective cubic is of the form

(52)
$$x^3 + ux + v = 0$$

and it is always possible to identify equations (51) and (52) by determining a and b in terms of u and v as roots of a quadratic equation. We find, indeed,

(53)
$$a^3 + b^3 = v, -3ab = u, a^3b^3 = -u^3/27$$

which means that a^3 and b^3 are roots of the equation

(54)
$$y^2 - vy - u^3/27 = 0$$

If we denote by A and B the roots of this *resolvent quadratic,* then $x = -\sqrt[3]{A} - \sqrt[3]{B}$ is a root of the cubic. By filling in the details of the calculations indicated, we arrive at the *Cardan formula for the cubic.*

Lagrange's solution of the defective quartic

(55)
$$x^4 + ux^2 + vx + w = 0$$

follows similar lines. The identity

(56) $(a + b + c + x)(a - b - c + x)(-a + b - c + x)$
              $(-a - b + c + x) = (a^4 + b^4 + c^4 + x^4) -$
              $2(a^2b^2 + b^2c^2 + c^2a^2 + x^2a^2 + x^2b^2 + x^2c^2) +$
              $8abcx = H(x)$

shows that

$$x = -a - b - c, -a + b + c, a - b + c, a + b - c$$

are roots of the defective quartic $H(x) = 0$.

Is it possible to determine a, b, c, in such a way that the two equations (55) and (56) be identical? Well, we have

$$a^2 + b^2 + c^2 = -u/2, \quad abc = v/8,$$
$$a^2b^2 + b^2c^2 + c^2d^2 = \frac{u^2 - 4w}{16}$$

from which we conclude that a^2, b^2, c^2, are roots of the equation

(57) $$y^3 + \frac{u}{2}y^2 + \frac{(u^2 - 4w)}{16}y - \frac{v}{8} = 0$$

If we denote by A, B, C, the roots of this *resolvent cubic,* then

$$x = -\sqrt{A} - \sqrt{B} - \sqrt{C}, \quad -\sqrt{A} + \sqrt{B} + \sqrt{C},$$
$$\sqrt{A} - \sqrt{B} + \sqrt{C}, \quad \sqrt{A} + \sqrt{B} - \sqrt{C}$$

are the roots of the quartic sought. Observe that the *radicands,* A, B, C, generally involve cubic surds.

This elegant approach suggests the question: Why couldn't a method, so successful in the case of cubics and quartics, be adapted to *quintic* and higher equations? Now, it is possible to set up some symmetric identity in *n* parameters and of degree *n* analogous to those used by Euler and Lagrange; by viewing one of these parameters as unknown, one could interpret the identity as an equation; by identifying the latter with the general equation of degree n, one could establish relations between the parameters and the coefficients, and, eventually, be led to an equation analogous to the *resolvents* of Euler and Lagrange.

As a matter of fact, several attempts in this direction were made, the most notable being that of Malfatti. The *Malfatti resolvent* for the general *quintic* is of the *sixth* degree, from which he conjectured that the degree *m* of the resolvent of an equation of degree *n* is given by the formula:

$$m = \tfrac{1}{2}(n - 1)(n - 2)$$

and that, consequently, for any value of n greater than 4, *the resolvent is of a higher degree than the original equation.*

Geometry and Graphics

Greek historians attribute the discovery of the *conic sections* to one Menaechmus, a contemporary of Plato and a disciple of Eudoxus, the mastermind who allegedly inspired Euclid's *Elements*. We are told that Menaechmus stumbled on these "solid loci" while in quest of a construction for *doubling the cube,* i.e., the solution of the equation $x^3 = 2$; and that he effectively solved the problem of a *parabola* and a circle. Details of his solution are not known, but a possible version of it is shown in Figure 4 which applies to the general *binomial cubic:* $x^3 - N = 0$. The equation of the parabola is $y = x^2$. The "resolvent" circle is erected on OP as diameter, P being the point: $x = N, y = 1$. The equation of the circle is therefore

(58) $$x^2 + y^2 - Nx - y = 0$$

Eliminating y between the two equations, we obtain: $x^4 - Nx = x(x^3 - N) = 0$. The parabola and the circle intersect in O and Q, and the absicssa of the latter is $x = -\sqrt[3]{A} - \sqrt[3]{B}$

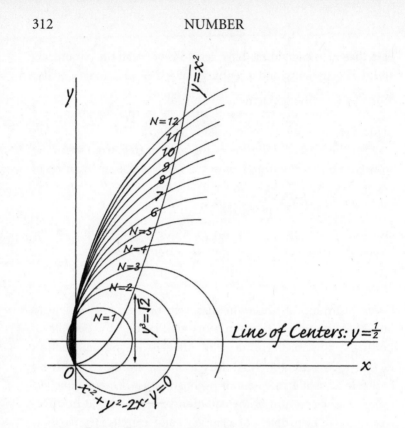

FIG. 4. GRAPHICAL EXTRACTION OF CUBE ROOTS

What lends considerable plausibility to the conjecture above is that about fifteen centuries later Omar Khayyám proposed a graphical solution of the general cubic and quartic based on the same principle, i.e., as *intersections of a fixed parabola and a mobile circle.* By eliminating y between

$$x^2 + y^2 + ux + vy + w = 0 \text{ and } y = x^2$$

we obtain the quartic equation

$$x^4 + (v + 1)x^2 + ux + w = 0$$

Let the *given quartic* be

$$x^4 + Ax^2 + Bx + C = 0$$

then by identifying coefficients we find

$$u = B, v = A - 1, w = C.$$

This completely determines the "resolvent" circle in terms of the coefficients of the given quartic. If the latter is a cubic, set $C = 0$, which leads to $w = 0$: in this case the *resolvent circle passes through the origin of coordinates*.

In Figure 5 the Omar method is applied to the *trisection of a general angle*. The given angle ϕ = xOM is represented by the point M on the *unit circle*, so that $a = \cos \phi = HC$. The *triple angle* formula

(59) $$\cos \phi = 4 \cos^3\phi/3 - 3 \cos \phi/3$$

leads to the cubic

(60) $$x^3 - 3x - 2a = 0$$

where $x = 2 \cos \phi/3$. The center C of the resolvent circle is in $x = a, y = 2$. Let Q be the point in the first quadrant common to this circle and the parabola; and let P be the projection of Q on the circle of center O and radius 2: then xOP is the angle sought.

The graphical solution of the quadratic equation

$$x^2 - ax + b = 0$$

where a and b are any real numbers, is shown in Figure 6. Here the resolvent circle is erected on UP as diameter, where P is the point of coordinates $x = a, y = b$, and U is the *unit-point* on the y-axis. The points X and X', where the resolvent circle crosses the x-axis, represent the graphical solution of the quadratic. If the circle does not intersect or touch the x-axis, then the roots are imaginary; still, in this case too the roots can be represented

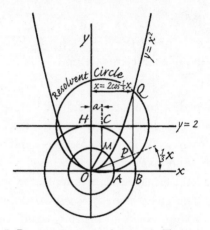

FIG. 5. PARABOLA AS AN ANGULAR TRISECTOR

graphically by means of a simple construction shown in Figure 7: Draw the tangent OT to the resolvent circle and let the circle which has O for center and OT for radius meet the line CM in Z and Z' mark the roots of the quadratic in the *plane of the complex variable* x + iy.

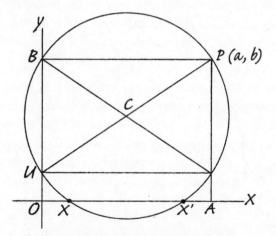

FIG. 6. GRAPHICAL SOLUTION OF A QUADRATIC EQUATION
WITH REAL ROOTS

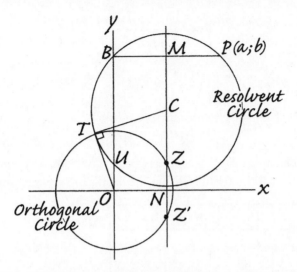

FIG. 7. GRAPHING THE COMPLEX ROOTS OF A QUADRATIC

The Euclidian Algorithm

The expansion of a number into a *continued fraction* is a variation of a procedure which appears in Book VII of the *Elements* and is, for this reason, called *Euclidean Algorithm.* Euclid used the method to determine the *greatest common divisor of two integers,* but the algorithm has many other applications. I shall discuss in this and following articles three such problems: *continued fractions, indeterminate equations,* and *irrational numbers.*

Let $[\Gamma]$ denote *the greatest integer contained in the positive number* Γ. Thus

$$[\tfrac{22}{7}] = 3, \quad [\sqrt{2}] = 1, \quad [\pi] = 3, \quad [e] = 2, \quad [1] = 1, \quad [0] = 0$$

Next assume Γ *rational.* If the number is an integer, then $\Gamma = [\Gamma]$; if Γ is not an integer, then there exists a rational number Γ_1 greater than 1 such that $\Gamma = [\Gamma] + (1/\Gamma_1) \cdot \Gamma_1$ may be an integer; if not, carry the algorithm to the next step; namely $\Gamma_1 = [\Gamma_1] + (1/\Gamma_2)$, and continue the process until an integer, say Γ_n is attained. In the end one arrives at the *continued fraction:*

(61)
$$\Gamma = [\Gamma] + \cfrac{1}{[\Gamma_1] + \cfrac{1}{[\Gamma_2] + \cfrac{1}{\ldots + \cfrac{1}{\Gamma_n}}}}$$

which I shall conveniently abridge into

(61') $\Gamma = ([\Gamma]; [\Gamma_1], [\Gamma_2], \ldots, \Gamma_n)$

As an example take $\Gamma = 106/39$ which is an *approximation of the number e* within .04%. Here are the details in tabulated form:

Dividend	Divisor	Quotient	Remainder
106	39	2	28
39	28	1	11
28	11	2	6
11	6	1	5
6	5	1	1
5	1	5	0

Hence
106/39 = (2; 1, 2, 1, 1, 5)

I call the array of integers $[\Gamma], [\Gamma_1], [\Gamma_2], \ldots, \Gamma_n$ the *spectrum of the number* Γ. The terms of the spectrum are the denominators of the regular continued fraction for Γ. A striking geometrical interpretation of the algorithm and of the spectra it generates is shown in Figure 8. We begin by erecting a rectangle of base 1 and height Γ; from this rectangle we *delete* as many squares as possible, leaving a *residual* rectangle, to which the same operation is applied; the process is continued until *no residual rectangle is left. The number of squares in any one of the successive rectangles gives the corresponding term in the spectrum of* Γ.

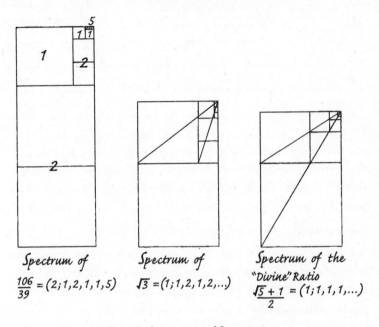

Spectrum of
$$\frac{106}{39} = (2;1,2,1,1,5)$$

Spectrum of
$$\sqrt{3} = (1;1,2,1,2,...)$$

Spectrum of the
"Divine" Ratio
$$\frac{\sqrt{5}+1}{2} = (1;1,1,1,...)$$

FIG. 8. SPECTRA OF NUMBERS

That the algorithm can be applied to any rational number, and that the resulting spectrum is *finite* and *unique* is inherent in the process. And, conversely, given any finite array of positive integers, there exists one and only one rational number, Γ, which admits the array for spectrum. A direct and arduous way of calculating Γ from its spectrum is to proceed "from bottom up." However, much labor can be saved by using another algorithm, the discovery of which is generally attributed to John Wallis, the teacher of Newton. To show how the procedure works in practice, I shall tabulate the results for the example treated above, $\Gamma = (2; 1, 2, 1, 1, 5)$, the first two *convergents* of which are, obviously, 2 and 3.

Term of Spectrum: g	2	1	2	1	1	5
Numerator of Convergent: N	2	3	$2 \cdot 3 + 2 = 8$	$1 \cdot 3 + 3 = 11$	19	106
Denominator of Convergent: D	1	1	$2 \cdot 1 + 1 = 8$	$1 \cdot 3 + 1 = 4$	7	39
Huygens Determinant: H		-1	$+1$	-1	$+1$	-1

Generally let N_{k-2}, N_{k-1}, N_k be the numerators, D_{k-2}, D_{k-1}, D_k be the denominators of *three consecutive convergents,* and let g_k be the term of the continued fraction which corresponds to the k-th convergent. Then, the Wallis algorithm is expressed in the *recursion law*:

(62) $N_k = g_k N_{k-1} + N_{k-2}$ $D_k = g_k D_{k-1} + D_{k-2},$

From these formulae Huygens deduced a theorem which later, at the hands of Euler and Lagrange, became the cornerstone in the theory of *infinite continued fractions.* Denote by H_k the determinant

(63) $H_k = \begin{vmatrix} N_{k-1} & N_k \\ D_{k-1} & D_k \end{vmatrix} = (-1)^k$

Then Huygen's theorem states that H_k *is alternately +1 or –1.*

On Indeterminate Equations

Many mathematical questions lead to the integral solutions of an algebraic equation in two or more unknowns. The study of Pythagorean triples, and the Fermat conjecture that the equation $x^n + y^n = z^n$ has no solution in integers, if $n>2$, belong to this class of problems which has received the general name of *Diophantine Analysis.*

The theme of the present article is the most elementary problem of Diophantine analysis: to determine all the integral solutions of the equation

(64) $qx - py = r$

where the integers, p, q, and r are *relatively prime in pairs.*

Observe first that if x = a, y = b is a solution of (64), then so is

$$x = a + np, \qquad y = b + nq$$

for *all integral values of n, positive or negative.* In the second place, if X and Y are solutions of the equation

(65) $$qX - pY = 1$$

then a = rX and b = rY satisfy equation (64). Finally, the equation

$$qx + py = r$$

can be brought to form (64) by the simple substitution x = x, y = −z.

Next, expand p/q into a continued fraction, and denote by N/D the *penultimate convergent* of the spectrum. Then, in virtue of Huygens' theorem, qN − pD is equal either to +1 or to −1. In the first case X = N, Y = D is a solution of equation (65), and in the second case X = −N, Y = −D is such a solution. It follows that the general solution of (64) is

(66) either $$x = np + rN, \qquad y = nq + rD$$
or $$x = np - rN, \qquad y = nq - rD$$

depending on whether the spectrum of p/q contains an *odd* or an *even* number of terms. *Example:* Consider the equation 39x − 106y = 5. The penultimate convergent of 106/39 is 19/7 and

$$\begin{vmatrix} 19 & 106 \\ 7 & 39 \end{vmatrix} = -1$$

Thus the general solution is x = 106n − 95, y = 39n − 35. Now, for n = 1 this yields the particular solution (11, 4); hence, we can also write

$$x = 106m + 11, y = 39m + 4,$$

where m can take on any integral value from −∞ to +∞.

A Problem in Cyclotomy

To erect a regular polygon of n sides is equivalent to dividing the circumference of a circle into n equal parts, of the construction of the angle $\omega = 2\pi/n$. This class of problems has received the name of *cyclotomy*. The last term is quite often used in the narrower sense of determining *the character of the integer n for which this construction can be effected by straightedge and compass exclusively*. Since the bisection of a circular arc is a ruler-compass operation, only the *odd values* of n are of interest.

The Greeks knew only three such integers, namely 3, 5, and 15. The construction of a regular *heptagon*, i.e., the division of a circle into 7 equal parts, was one of the celebrated problems of antiquity. It was not solved until the eighteenth century, when it was established that the cases n = 7 and n = 13 lead to *irreducible cubics*.

Such was the status of the matter in 1801, when the problem was reformulated and completely solved by Gauss. His theorem states that if it is possible to divide the circumference into n equal parts by means of straightedge and compass, then n is either a *Fermat prime* or a *square-free product of such primes*. Let us recall that a Fermat prime is of the form $2^{2^P} + 1$. In virtue of the Gauss theorem there are 8 "cyclotomic" odd integers on this side of 300:

(67)   3, 5, 15, 17, 51, 85, 255, and 257

The first part of Gauss' theorem implies the study of the equation $z^n - 1 = 0$, and is not within the scope of this work. The second part is equivalent to the statement that if it is possible to divide a circle into p parts and also into q parts, then *it is possible to divide it into pq parts*, provided p and q are *relatively prime*. Let us set

(68) $2\pi/p = \alpha$, $2\pi/q = \beta$, $2\pi/pq = \gamma$

We assume that we can construct the arcs α and β by the means at hand; by he same token we could construct the arcs xα and yβ where x and y are any integers whatsoever, and consequently the arc xα – yβ, inasmuch as such operations can be executed by the compass only. The question is to find two integers x and y which satisfy the relation

$$\gamma = x\alpha - y\beta$$

Now, in virtue of (68) this last relation is equivalent to the *indeterminate equation.*

$$xq - yp = 1$$

which, as we saw in the preceding article, has always solutions *as long as p and q are relatively prime.*

As an example, let p = 51, q = 40. By Euclid's algorithm we find 51/40 = (1; 3, 1, 1, 1, 3), the penultimate convergent of which is 14/11. Accordingly

$$11\left(\frac{2\pi}{40}\right) - 14\left(\frac{2\pi}{51}\right) = \frac{2\pi}{40 \cdot 51}$$

Infinite Spectra

There is nothing either in the definition or in the execution of the Euclidean algorithm that should restrict it to rational numbers, except that when applied to irrational magnitudes the process is always *interminate.* Nor is the Wallis' algorithm restricted to a finite spectrum. Indeed, in virtue of Huygens' theorem, *the algorithm converges for any infinite spectrum,* or, in more precise terms, *the convergents of any regular infinite continued fraction approach some unique irrational number as a limit.* The question whether the application of the Euclidean algorithm to this irrational always re-creates the original spectrum raises some fine points which I cannot discuss here. However, the answer is in the affirmative.

Thus, the two algorithms when used in conjunction constitute a most powerful tool for deriving *rational approximations to irrational magnitudes.* The practical aspect of this problem involves a term-by-term expansion of the number into a regulator continued fraction, and this is a matter of laborious routine at worst. But to uncover in the infinite spectrum some mathematical pattern or some law of succession is quite another story. Indeed, apart from quadratic expressions, the spectra of irrational magnitudes, whether algebraic or transcendental, appear generally as formless and inarticulate, in short, as *random* as the decimal expansion of π.

By contrast, the spectra of irrationals of the type $A + \sqrt{B}$ are anything but random, and this is particularly true of the quadratic surds of integers. The *periods*, the patterns of the *cycles*, the number-theoretical properties of the convergents, the intimate connection between these convergents and some classical Diophantine problems have been sources of great fascination to many mathematicians of the eighteenth and nineteenth centuries, from Euler to Sylvester, and have not lost their interest to this day. Lack of space prevents me from presenting this development in a comprehensive form; the examples below have been selected in the hope that the reader may be stimulated to a further study of the subject.

Surds and Cycles

We begin with examining the operation of the Euclidean algorithm on a typical surd such as $\sqrt{23}$. Accordingly , let $N = 23 = 4^2 + 7$. The first term of the spectrum of $\sqrt{23}$ is, therefore, 4; the *first residue* is $\sqrt{23} - 4$; its reciprocal $(\sqrt{23} + 4)/7$, is the second *complete quotient;* and the greatest integer contained in this quotient is the second term of the spectrum. Continuing in this manner, we eventually arrive at a residue equal to the first, and from this step on the terms of the spectrum begin to repeat. Thus

$$\sqrt{23} = (4;\ 1, 3, 1, 8,\qquad 1, 3, 1, 8,\qquad 1, 3, 1, 8, \ldots)$$

which may be conveniently abridged into $(4;\ \underline{1, 3, 1, 8})$. The set 1, 3, 1, 8 is called the cycle of the spectrum, the number of terms in the cycle is the *period*. Details of the algorithm are shown in Table I. Table II lists the cycles of surds from $\sqrt{2}$ to $\sqrt{24}$.

TABLE I. TYPICAL EXPANSION

Complete Quotient	$\sqrt{23}$	$\dfrac{\sqrt{23}+4}{7}$	$\dfrac{\sqrt{23}+3}{2}$	$\dfrac{\sqrt{23}+3}{7}$	$\sqrt{23}+4$	$\dfrac{\sqrt{23}+4}{7}$
Term of Spectrum	4	1	3	1	8	1
Residue	$\sqrt{23}-4$	$\dfrac{\sqrt{23}-3}{7}$	$\dfrac{\sqrt{23}-3}{2}$	$\dfrac{\sqrt{23}-4}{7}$	$\sqrt{23}-4$	$\dfrac{\sqrt{23}-3}{7}$

Passing to the general case, let N be any non-square integer. Then we can write $N = b^2 + h$, where h may range from 1 to 2b. Let us also set $[2b/h] = c$. With this notation the general properties of the expansion are:

1) The spectrum of $\sqrt{b^2+h}$ begins with b and c;

2) c is the *first term of the cycle* and also its *penultimate* term; the *last term* of the cycle is 2b;

3) the terms of the cycle which precede 2b form a *symmetric block*. If the period, p, is even, as in the case of $\sqrt{14}$ or $\sqrt{23}$, then the *symmetry is odd*, and there is *one central term*. If the period is *odd*, as in $\sqrt{13}$, then the *symmetry is even*, and there are *two central terms*.

Thus the condition that a periodic spectrum represent the square root of an integer is that it be of the form

$$\Gamma = (b; \ \overline{c, d, \ldots d, c, 2b})$$

However, we shall presently see that while this *symmetry is necessary, it is not sufficient.*

TABLE II. CYCLES OF SURDS

Surd	b	middle	2b	Surd	b	middle	2b
$\sqrt{2}$	1		2	$\sqrt{14}$	3	1 2 1	6
$\sqrt{3}$	1	1	2	$\sqrt{15}$	3	1	6
$\sqrt{5}$	2		4	$\sqrt{17}$	4		8
$\sqrt{6}$	2	2	4	$\sqrt{18}$	4	4	8
$\sqrt{7}$	2	1 1 1	4	$\sqrt{19}$	4	2 1 3 1 2	8
$\sqrt{8}$	2	2	4	$\sqrt{20}$	4	2	8
$\sqrt{10}$	3		6	$\sqrt{21}$	4	1 1 2 1 1	8
$\sqrt{11}$	3	3	6	$\sqrt{22}$	4	1 2 4 2 1	8
$\sqrt{12}$	3	2	6	$\sqrt{23}$	4	1 3 1	8
$\sqrt{13}$	3	1 1 1 1	6	$\sqrt{24}$	4	1	8

Of special interest are the cyclic continued fractions of *period 2*. If such a fraction is to represent the square root of an integer, then it must be of the form

$$x = (b; \overline{c, 2b}), \text{ i.e. } \quad x - b = \cfrac{1}{c + \cfrac{1}{x + b}}$$

The last relation leads to the quadratic

$$x^2 = b^2 + 2b/c$$

It follows that in order that $(b; \overline{c, 2b})$ represent the square root of an integer, c *must divide 2b*. This will certainly occur when $c = 1$ or 2, or b or 2b; which leads to

$$\sqrt{b^2 + 2b} = (b; \overline{1, 2b}) \qquad \sqrt{b^2 + b} = (b; \overline{2, 2b})$$

$$\sqrt{b^2 + 2} = (b; \overline{b, 2b}) \qquad \sqrt{b^2 + 1} = (b; \overline{2b})$$

In this last case, the period drops to 1. If b is a *prime*, these four are the only spectra of periods 1 or 2. It is otherwise when b is a composite integer, such as 30, for instance. Here $2b = 60$, and c may be 1, 2, 3, 4, 5, 6, 10, 12, 15, 20, 30, 60. It follows that between $\sqrt{901}$ and $\sqrt{961}$ there are 12 surds for which the period is less than 3.

On Principles and Arguments

"Mathematicians do not deal in objects, but in relations between objects; thus, they are free toreplace some objects by others so long as the relations remain unchanged. Content to them is irrelevant: they are interested in form only."

—Poincaré

On Dirichlet's Distribution Principle

If a chest of five drawers contains more than five shirts then one of the drawers contains more than one shirt. If a family consists of more than seven members, then at least two of the family were born on the same day of the week. If there are more trees in a forest than leaves on any one tree, then at least two of the trees have the same number of leaves. These are examples of an argument which has come to be known as the *Dirichlet Box Principle*. Expressed in general terms, the principle states that *if p objects occupy q compartments, and p is greater than q, then at least one compartment contains more than one object.*

I give no proof of this proposition here for fear that readers may accuse me of "accentuating the obvious." Such, at least, was the reaction of a lay friend who heard the argument for the first time. "It is evident to the dullest mind," he said, "that if there are more guests at a luncheon than rolls, then some rolls will have to be divided, or some guests will go without rolls. Then why call such a truism a principle, and name it after a

327

mathematician, and a nineteenth century mathematician at that? How can an idea so basic and simple be viewed as a modern discovery, when it must have been implicit in mathematical reasoning since its inception?"

These remarks are well taken. Yet they would apply with equal force to other devices which had been implicitly used by mathematicians and laymen alike long before they were formulated as principles. Examples of such *latent ideas* are: Pascal's principle of *Complete Induction*, Fermat's method of *Indefinite Descent*, the *Dedekind Cut*. But why stop here? *Deductive Inference* was the medium of theological speculation ages before Thales made it the *sine qua non* of mathematical reasoning; while the *principle of position* has been an organic aspect of human speech ever since man has felt the necessity of expressing in words numbers which exceeded his finger aggregate.

However, it is one thing to use a mental device as a daily routine, quite another to express the idea in explicit terms, and then apply it consciously to the exploration of a new field of thought. Remember the parvenu in Molière's play who discovered to his consternation that he had been talking *prose* all his life? Well, we all use prose, and most of us abuse it; still, prose remains the most powerful medium for expressing thought and conveying it to posterity.

The distribution principle has been tacitly invoked in this work on several occasions: the determination of the period of the decimal expansion of a rational fraction; the Gaussian proof of Wilson's theorem; the expansion of a quadratic surd into a regular continued fraction—to mention but a few. I shall review here the first, because it is typical of the way the principle operates as an instrument of proof.

Consider the infinite decimal fraction representing 1/q, where q is any integer not divisible by 2 or 5. The expansion is *cyclic,* and its period p, i.e., the number of digits in a cycle, is a function of q, the precise nature of which has not been determined to this day. We do know, however, that p cannot exceed q − 1. Indeed, p is equal to the length of the residue cycle (see A8), the elements of which may range from 1 to q − 1. If the period p exceeded q − 1, then according to Dirichlet's principle the same residue would occur more than once in the same cycle, which would contradict the fact that no two residues of a cycle can be equal.

The Terms "Possible" and "Impossible" in Geometry

The difficulties incident to such problems as doubling the cube, trisecting an angle, squaring the circle, and others bequeathed to us by the ancients were not inherent in the problems themselves. The difficulties merely reflected the drastic character of the *classical interdiction* which confined geometrical construction to manipulations by straightedge and compass.

The terms "possible" or "impossible," as applied to geometrical construction, have no *absolute* significance: we must stipulate in each case the equipment by means of which the construction is to be executed. Indeed, were all restrictions removed, were any device susceptible of geometrical formulation admitted on equal terms with the traditional instruments, were any geometrical locus, whether generated mechanically or by some graphical procedure, accepted at par with the line and the circle—the words "possible" or "impossible" would lose all meaning, for obviously the field of *possible* problems would become co-extensive with the field of *all* problems.

In the classical treatment of construction, the instrument remained modestly in the background; the modern approach puts it prominently to the fore. Each instrument is viewed as the "exponent" of a whole group of problems which may be said to constitute its domain. Thus we have the *linear* domain, made up of problems which could be solved by the straightedge alone; the *circular* domain, or domain of the compass; and the *classical*, or traditional domain, which comprises both the linear and circular. Beyond the traditional domain lies the vast territory of problems for which the classical instruments do not suffice.

On the Diagonal Process

We wish to prove that *the aggregate of real numbers is not denumerable.* To this end we assume that any one of the real numbers in the interval $0 < x < 1$ has been expressed as a decimal fraction. Strict adherence to this procedure would lead to ambiguity, such as .5 and .4999 ... representing the same rational number. To avoid this difficulty we agree to replace every terminating decimal fraction by its non-terminating equivalent, as in the example above.

Let us assume next that one could enumerate the aggregate of real numbers: then it would be possible to arrange them into a sequence of the type

$$x_1 = 0.a_1 \ a_2 \ a_3 \ ...$$

$$x_2 = 0.b_1 \ b_2 \ b_3 \ ...$$

$$x_3 = 0.c_1 \ c_2 \ c_3 \ ...$$

$$...\qquad\quad...\qquad\quad...$$

We are going to show that regardless of how this array has been constructed, there always exists a real number x' which does not enter therein. x' is defined by an algorithm which has received the name of *diagonal process.*

$$x' = 0.a_1' \ b_2' \ c_3' \ ...$$

where a_1' is a digit which is neither 0 nor a_1; similarly, b_2' is neither 0 nor b_2, etc.

In this manner we have defined x' by means of a decimal fraction with an infinite number of significant digits, and yet this real number is different from any of the numbers in the sequence above, because two decimal fractions with an infinite number of significant digits are equal if, and only if, they are identical digit by digit; but x' differs from x_1 in the first digit, from x_2 in the second, and in general, from x_n in the n-th digit. We have thus proved the existence of a real number x̄ not contained in the sequence above, which contradicts the assumption that the aggregate of real numbers is denumerable.

On Bounded Geometry

Empirical geometry is, in its very nature, a *bounded* geometry, and to establish general laws for such a geometry would be a matter of extreme difficulty. Think of your desk, for instance, as an exclusive field of geometrical speculation. You will have to distinguish between a great variety of straight lines: there are those that issue from the corners of your desk; there are others that intercept two diagonals; there are the lines that meet the edges at right angles, etc. Take four points at random, and ask whether the line joining the first two points would meet the line adjoining the remaining two: you would have to ascertain first to which of the several species one or the other line belonged. If, furthermore, you should attempt to analyze all possible cases with a view to establishing criteria for the intersection of two lines, you would be led to formulate a law in comparison with which the rules for forming the past of English irregular verbs would appear as child's play.

In a *bounded geometry*, such problems as the construction of a circle with a given radius and a given center, or the circumscribing of a circle about a triangle, or the dropping of a perpendicular from a point to a line would generally have no solution. Two lines would generally not form an angle, nor three a triangle. To stipulate a triangle similar to a given triangle would be meaningless if the scale exceeded a certain number. The shortest distance from a point to a line would not always be the perpendicular, etc.

But even had you succeeded in mastering this intricate geometry, your difficulties would be just beginning. For were the boundary of your geometrical field changed from the rectangle of your desk to some other contour, say to a triangle, circle, or oval, you would have to start anew. It is indeed characteristic of a bounded geometry that its laws depend essentially on the nature of the boundary within which it is being practiced. Here again we may utilize the analogy with language: the rules of a bounded geometry would be like the grammar of an individual language: every boundary would require a distinct set of such rules. And, while some of these rules would be common to all boundaries, the salient exceptions would be as different for two distinct boundaries as are English and French irregular verbs. And if this is true of the empirical geometry of a plane boundary, what difficulties would arise in connection with space configurations?

It is obvious enough, therefore, that had we been confined to a finite boundary, the deductive method would have been of little avail: geometry would have remained a descriptive science, attaining no greater degree of generality than zoology, botany, or mineralogy.

On the Principle of Transitivity

If a relation is such that when it holds between A and B, and between B and C, it also holds between A and C, then we say that it is *transitive*. To illustrate: blood relationship is a *transitive* relation, parenthood is *intransitive*. Examples of transitivity in mathematics are: *equality, congruence, parallelism;* of intransitivity—

the relation of *overlapping*, also of *"in-and-out."* Thus, figure A may be inscribed in figure B, and B inscribed in C, without A being inscribed in C.

The *principle of transitivity* consists in the statement that *if two things are in some way equivalent to a third they are equivalent to each other*. This principle is of the greatest practical importance in many questions. Thus in geometry we define two segments as *congruent* if it is possible so to displace one as to make it coincident with the other: if we took this criterion seriously, we would have to cut out a portion of the plane which contains one of the two segments and place it on another portion of the plane. In practice, of course, we do nothing of the kind: in practice we use the compass, a graduated ruler, or divisors, invoking in each case that *two segments congruent to a third are congruent to each other*. In pure mathematics the principle of transitivity is invoked whenever we transform an equality from one form to another. In short, *the most fundamental aspect of mathematical equality is its transitivity*. How about *physical equality?*

To bring out the issue as concretely as possible, I shall ask the reader to imagine that he has been presented with a number of steel bars, identical except for their varying lengths. To be specific, let us assume that these have been carefully measured in the laboratory and found to range from 30 to 50 millimeters; in particular, three of these bars, marked A, B, and C, measure 30, 31, and 32 millimeters respectively. Of this, however, you know nothing; nor do you want to know, since this information may prejudice your judgment; for you aim at ascertaining what sort of measuring technique you could develop with your sense, unaided by instruments and gauges.

You commence by laying the bars A and B side by side: you find that neither your eye nor your fingertips can discern any difference between the lengths of these bars; so you declare them *identical.* You repeat the same comparative test with B and C; you decide that these bars, too, are identical in length. Next you juxtapose A and C; but now both your eyes and your fingertips unmistakably discern that C is longer than A. You arrive at the startling conclusion that *two things may be equal to a third without being equal to each other.*

But this conclusion stands in direct contradiction with one of the most important axioms of mathematics, which asserts that two quantities equal to a third are necessarily equal to each other. This axiom is back of most of the operations of arithmetic; without it we could neither transform identities nor solve equations. I should not go so far as to say that a mathematics denying this axiom could not be constructed. The important fact is that the physicist uses no such *modernistic* discipline, but the *classical mathematics* of which this axiom is a cornerstone.

What gives him the right to do it? Could it be that the introduction of scientific measuring devices in lieu of direct perception has removed the contradiction? No. *Reading a graduated scale* is the ultimate goal of any measuring device; consequently, however ingenious may be the designer of the instrument, he must, in the last analysis, rely on the sense of some observer, more particularly on his vision. When, on the other hand, we examine more closely the operation of reading a scale, we find that it does not differ in any *essential* feature from the hypothetical case of the bars considered above. To be sure, the *critical interval,* which in that case was one millimeter, may now have been contracted to one micron; through amplification, and by rendering the measuring device more sensitive by various precision methods, one may even succeed in reducing the interval to a small fraction of a micron. And yet it is obvious enough that no matter how far this process of refinement be carried, it cannot eliminate our difficulty, nor even minimize it; for in the end data must remain of which one could say: "I find measure A identical with measure B; I also find measure C to be identical with B; still I can distinctly discern that C is greater than A."

The Measurable and the Commensurate

There is ample evidence that the ancients were aware of the existence of rational triangles, such as are given by the triples (3, 4, 5), (5, 12, 13), and (7, 24, 25); and it was, undoubtedly, the search for additional triples that had led the early Greek mathematicians to the Pythagorean theorem.

The latter was a triumphant confirmation of their number philosophy. However, the triumph was short-lived: for the very generality of the proposition revealed the existence of *irrational magnitudes*. One effect of this perturbing discovery was a revised outlook on matters of geometry. To the early Pythagoreans *every triangle was a rational triangle*, because they held that *all things measurable were commensurate*. This last dictum seems to them as incontrovertible as any axiom; and when they proclaimed that number *ruled the universe*, they meant by number *integer*, for the very conception that magnitudes might exist which were not directly amenable to integers was alien to their outlook as well as to their experience.

Some modern interpreters of mathematical thought have been inclined to dismiss the ideas of the early Pythagoreans as naïve notions of a bygone age. And yet in the eyes of the individual who uses mathematical tools in his daily work—and his name today is legion—but to whom mathematics is but a means to an end, and never an end in itself, these notions are neither obsolete nor naïve. For such numbers as are of practical significance to him result either from *counting* or from *measuring*, and are, therefore, either *integers* or *rational fractions.* To be sure, he may have learned to use with comparative facility symbols and terms which allude to the existence of non-rational entities, but this phraseology is to him but a useful turn of speech. In the end, the rational number emerges as the only magnitude that can be put to practical use.

Should this individual, piqued by the reproach that he was naïve, endeavor to penetrate behind the mysterious nomenclature, he would soon discover that the processes invoked to vindicate these non-rational beings are wholly unattainable and, therefore, gratuitous. Should he persist, should he attempt to interpret these entities in his own rational terms, he would be sternly reminded that *in matters irrational one may at times evade the infinite, but never avoid it.* For inherent in the very nature of this mysterious being is the property that no matter how close any given rational number may "resemble" it, other rational numbers exist which "resemble" it even more closely.

This individual would feel far more at home among the early Pythagoreans than among their more rigorous successors. He would willingly embrace their credo that all things measurable are commensurate. Indeed, he would be at a loss to understand why a principle so beautiful in its simplicity was so wantonly dismissed. And, in the end, *the mathematician would be forced to concede that the principle was abandoned not because it contradicted experience, but because it was found to be incompatible with the axioms of geometry.*

For if the axioms of geometry are valid, then the Pythagorean theorem holds without exceptions. And if the theorem holds, then the square erected on the diagonal of a square of side 1 has an area equal to 2. If, on the other hand, the Pythagorean dictum held, then 2 would be the square of some rational number, in flagrant contradiction to the axioms of arithmetic.

Time and the Continuum

Your consciousness attests of the *now;* your mind recalls other *nows,* less distinct as they recede into the *past* until lost in the hazy dawn of memory. These temporal series, vague and over-lapping, you attach to one individual, whom you call *I.* In the course of a few years, every cell of this individual's body has changed; his thoughts, judgments, emotions, and aspirations have undergone a similar metamorphosis. What is then this per-manence which you designate as *I?* Surely not the mere name which differentiates this individual from his fellow men! Is it then this temporal series strung like beads on the filament of memory?

A discrete sequence of disjointed recollections which begins some time in infancy and abruptly terminates with the present, such is *time* as an immediate datum of consciousness. When, however, this raw material has undergone the mysterious refin-ing process known as physical intuition, it emerges as something quite different. *Intuitive time is extrapolated time,* extrapolated beyond the dawn of consciousness, into the infinite recesses of the *past,* and beyond the present into the infinite *future,* this lat-ter also being conceived as made up of *nows,* as the past has been. By an act of our mind we separate time into these two classes, the past and the future, which are mutually exclusive and together comprise *all* of time, *eternity.* The *now* to our mind is but a partition which separates the past from the future; and since any instant of the past was once a *now,* and since any instant of the future will be a *now* anon, we conceive any instant of the past or the future as such a partition.

Is this all? No, intuitive time is *interpolated* time: between any two instants of the past, however closely associated in our memory, we insert—again by an act of the mind—other instants, in number indefinite. This is what we mean by *conti-nuity* of the past; and the same continuity we impose on the

future. *Time, to our mind is a stream;* to be sure, of this stream
our experience knows but disconnected elements; yet our intuition fills in the gaps left by experience; it converts time into a
continuum, the prototype of all continua in nature.

What is, for instance, that perfect continuity which we
ascribe to a geometrical line, if not the conviction that we can
describe such a line by an *uninterrupted* motion of the hand? We
transfer the streamlike character of duration to all physical phenomena our first attempt to analyze any phenomenon, whether
it be light or sound, heat or electricity, is to express it in terms of
distance, mass, or energy, so that we may reduce it to a *function
of time.*

The conflict between the discrete and the continuous is not
a mere product of school dialectics: it may be traced to the very
origin of thought, for it is but the reflection of the ever present
discord between this conception of time as a stream and the discontinuous character of all experience. For, in the ultimate analysis, our number concept rests on *counting,* i.e., on enumerating the
discrete, discontinuous, interrupted, while our time intuition
paints all phenomena as flowing. To reduce a physical phenomenon to number without destroying its streamlike character—
such is the Herculean task of the mathematical physicist; and, in
a broad sense, geometry too should be viewed as but a branch of
physics.

Mathematics and Reality

Classical science assigned to man an exceptional position in the
scheme of things: he was capable of detaching himself from the
ties which chained him to the universal mechanism, and of
appraising this latter in true perspective. To be sure, his consciousness too was viewed as a link in this endless chain of cause
and effect, yet the evolution of this consciousness was believed
to be in the direction of greater *freedom.* His body was chained,

but his mind was free to contemplate these chains, to classify, measure, and weigh them. The book of nature lay open before his eyes; he had but to decipher the code in which it was written, and his faculties were equal to the task.

This code was rational: the immutable order that was man's to contemplate was governed by rational laws; the universe had been designed on patterns which human reason would have devised, had it been entrusted with the task; the structure of the universe was reducible to a rational discipline; its code of laws could be deduced from a finite body of premises by means of the syllogisms of formal logic. These premises derived their validity not from speculation but from experience, which alone could decide the merit of a theory. Like Antaeus, who, harassed by Hercules, would restore his waning strength every time his body touched his mother Earth, so did speculation constantly gain by contact with the firm reality of experience.

The mathematical method reflected the universe. It had the power to produce an inexhaustible variety of rational forms. Among these was that cosmic form which some day may embrace the universe in a single sweep. By successive approximations science would eventually attain this cosmic form, for with each successive step it was getting nearer and nearer to it. The very structure of mathematics guaranteed this *asymptotic* approach, since every successive generalization embraced a larger portion of the universe, without ever surrendering any of the previously acquired territory.

Mathematics and experiment reign more firmly than ever over the new physics, but an all-pervading skepticism has affected their validity. Man's confident belief in the absolute validity of the two methods has been found to be of an anthropomorphic origin; both have been found to rest on articles of faith.

Mathematics would collapse like a house of cards were it deprived of the certainties that man may safely proceed as though he possessed an unlimited memory, and an inexhaustible life lay ahead of him. It is on this assumption that the validity of infinite processes is based, and these processes dominate mathematical analysis. But this is not all: arithmetic itself would lose its generality were this hypothesis refuted, for our concept of whole number is inseparable from it; and so would geometry and mechanics. This catastrophe would in turn uproot the whole edifice of the physical sciences.

The validity of experience rests on our faith that the future will resemble the past. We believe that because in a series of events which appear to us similar in character a certain tendency has manifested itself, this tendency reveals permanence, and that this permanence will be the more assured for the future, the more uniformly and regularly it has been witnessed in the past. And yet this validity of *inference,* on which all empirical knowledge is based, may rest on no firmer foundation than the human longing for certainty and permanence.

And this unbridgeable chasm between our unorganized experience and systematic experiment! Our instruments of detection and measurement, which we have been trained to regard as refined extensions of our senses, are they not like loaded dice, charged as they are with preconceived notions concerning the very things which we are seeking to determine? Is not our scientific knowledge a colossal, even though unconscious, attempt to counterfeit by number the vague and elusive world disclosed to our senses? Color, sound, and warmth reduced to frequencies of vibrations, taste and odor to numerical subscripts in chemical formulae, are these the reality that pervades our consciousness?

In this, then, modern science differs from its classical predecessor: it has recognized the anthropomorphic origin and nature of human knowledge. Be it determinism or rationality, empiricism or the mathematical method, it has recognized that *man is the measure of all things, and that there is no other measure.*

The End.

Afterword

Since the fourth and last edition of *Number* was published, a half-century ago, mathematics has advanced with astonishing speed. Several of the most outstanding unsolved problems have either been solved or spread roots to new places in nearby fields. From the time our ancestors first discovered rules for operating with numbers, problems of mathematics cropped up; some were solved, others not; but, like stones in ancient Phoenician barley fields, new ones surfaced faster than the old were removed. Yet, despite developments in modern number theory and analysis, the content of *Number* is still as fresh as when the first edition was published in 1930. Reading *Number* today, the mathematics enthusiast is struck by its lucid language, contemporary relevance, and intellectual provocation.

Progress in mathematics has accelerated. On the surface, it may seem as if only a few famous problems have been solved in the past 50 years. But modern mathematics has increasingly become more profound. Solutions to surface problems—the so-called "gems"—are inextricably linked to others that are often fields apart, crossing boundaries by intricately tangled roots coming from one great and stable unifying source.

The ancient problems of doubling of the cube, trisecting the angle, and squaring the circle remained a mystery for two thousand years, waiting for the brilliant ideas of modern algebra to uncover their proofs. In 1837, Pierre Wantzel proved that it is impossible to duplicate the cube or trisect an arbitrary angle, thereby solving the two great mysteries of antiquity. Was that the end of the long story that began with the tale of the oracle at Delos, which claimed that relief of the devastating plague in Athens would come when the cubic altar to Apollo would be doubled in size? Certainly not! Wantzel's solution opened new questions, questions on which simple algebraic criteria would permit geometric constructions as solutions of rational polynomial equations. These questions, in turn, opened the far broader question of how to convert geometry to the theory of equations.

Dantzig focused on the evolution of the number concept to keep his book well within a manageable scope, staying reasonably clear of the more geometric branches of mathematics, even though he knew that answers to some of the most elementary questions of number theory are sometimes best handled through sophisticated geometry. His book mentions the Goldbach Conjecture, the Twin Prime Conjecture, Fermat's Last Theorem; three of many outstanding statements still unproven at the time of its last printing. Fermat's Last Theorem was solved in 1994 by Andrew Wiles, with the help of his former student Richard Taylor, using some of the most beautiful and brilliant ideas in number theory that recognize relationships between outwardly different mathematical objects coming from remotely different branches of mathematics. (I cannot presume to give anything near an adequate story here because the formal proof is highly technical, but it has been comprehensively outlined in several popular books listed in the Further Readings section.) The other two conjectures remain unsolved.

The Twin Prime Conjecture, for example, is one of a large assortment of problems prompted by asking simple, phenomenological questions about how the collection of prime numbers is distributed among all natural numbers. The wonderful thing about many of the finest questions in number theory is that they can be stated so simply. They require little or no technical language to understand and can often attract the least-suspecting visitor, who—if not careful—may find him- or herself absorbed in endless hours of mathematical diversions. How many prime numbers are there of the form $n^2 + 1$? How many prime numbers p are there with $2p + 1$ being a prime number? Are there any odd perfect numbers? (Perfect numbers, such as 6, are equal to the sum of their own divisors.) We now know that there are none under 300 digits. But are there any? We know that if one exists at all it must be a sum of squares and at the same time have at least 47 prime factors. But are there any at all?

There was a time when young, naïve mathematicians (like myself) would worry about what would happen when all these fine questions—those simply stated ones—would be solved. We have learned not to worry. Not only will there always be enough fine questions to tempt the dilettante, but each answer will breed a family of new ones. Such was certainly the case with Fermat's Last Theorem, which reared much of modern number theory; it was also the case with those stubborn ancient Greek problems, which formed so much of modern algebra. We forever find ourselves at those relatively earlier stages of understanding number.

Fifty years may seem like a long time to wait for solutions to outstanding problems, but considering that some have waited millennia it seems that plenty has happened in the mere 2 percent of the time since Euclid's *Elements* first appeared and modern mathematics took off. First, we'll look at how computers have affected mathematics. Then we'll take a peek at the progress on the Goldbach Conjecture and the Twin Prime Conjecture.

Computers

In 1954, the year the fourth edition of *Number* was published, MANIAC I (Mathematical Analyzer, Numerical Integrator and Computer) was the most advanced computer of the time, using 18,000 vacuum tubes. (One can only imagine how often the machine broke down because a single 1 of the 18,000 tubes failed.) In 1951, without the use of computers, the 44-digit number

$$(2^{148} + 1)/17 = 20988936657440586486151264256610222593863921$$

was discovered as the largest prime, but just three years later, with the help of MANIAC I, the largest prime was discovered to be $2^{2,281} - 1$, a number with 687 digits. Today we know that $2^{24,036,583} - 1$ is a prime number. It contains 7,235,733 digits.

In 1954, graphics interface analogue printers were still on the drawing boards, although prototypes that moved styluses up, down, right, or left according to the coordinates of input were being built by IBM. Dantzig does not mention the Riemann-Zeta function, but the zeros of that interesting function (solutions to the equation $\zeta(s) = 0$) have a curious connection with the distribution of prime numbers. A flood of number theory theorems would automatically follow from a proof of the Riemann Hypothesis, which claims that all the zeros of $\zeta(s)$ are complex numbers of the form $1/2 + ai$. For one, in 1962 Wang Yuan showed that if the Riemann Hypothesis is true, then there are infinitely many primes p such that p and $p + 2$ are a product of at most three primes. Riemann was able to compute the first three zeros of the zeta function with astonishing accuracy by hand. In 1954, when Alan Turing found 1,054 zeros of the zeta function without an electronic computer, 1,054 seemed like a huge number of zeros; but now, with the aid of modern computers, we know more than 10^{22} zeros and all of them are on the line having its real part equal to 1/2. Today, the world's fastest computer cannot possibly tell whether *all* zeros of the Riemann-Zeta function lie on the vertical line $1/2 + ai$ in the complex plane, but a simple $500 desktop computer can instantly find many that do, yet never find any that do not.

But computers work with finite numbers and although they can work at astonishing speeds, those speeds are only finite. They can help discovery, relieve the mathematician of grueling endless computations, and—in many cases—suggest possibilities that could never have been spotted by human reckoning.

The Goldbach Conjecture

We now know a few things about the Goldbach Conjecture, which says that every even number greater than 2 can be written as a sum of two primes. Dantzig knew, but didn't mention, that every sufficiently large odd number can be written as a sum of three primes. The Russian mathematician Ivan Vinogradov proved this in 1937. Dantzig also knew the wild but interesting theorem that claimed that every positive integer could be written as the sum of not more than 300,000 primes. Now that may seem like a long way off from Goldbach's Conjecture, but in fact 300,000 is a lot less than infinity! Lev Shnirelmann, another Russian, proved it in 1931. Soon after, Vinogradov used methods of Hardy, Littlewood, and Ramanujan to prove that any sufficiently large number could be written as a sum of four primes. In more precise terms, it means that there exists some number N such that any integer greater than N can be written as a sum of four primes. This brought down the number of primes in the sum at the expense of the size of the number for which the conjecture would be true.

Vinogradov proved both theorems by exhibiting a contradiction from the assumption that infinitely many integers cannot be written as a sum of four primes. His proof could not specify how large N had to be, but in 1956, K. G. Borodzkin showed that N had only to be greater than $10^{4,00,8,660}$, a number with more than four million digits. It is now known that "almost all" even numbers can be written as the sum of two primes. "Almost all" here means that the percentage of even numbers under N for which the Goldbach Conjectures are true tends toward 100 as N grows large. Just after the last printing of *Number*, there was a flurry of theorems closing in on the classical Goldbach Conjecture. First, it was proven that every sufficiently large even integer is the sum of a prime and a product of at most nine primes. As the years went by, the product was reduced, first to five, then to four, then to three, and finally to two. We now know that every sufficiently large even integer is the sum of a prime and the product of two primes. We also now know that one Goldbach variation is true: With a finite number of exceptions, every even number is a sum of a pair of twin primes.

Twin Primes

It is still not known whether there are an infinite number of twin primes, but it seems certain that there are. Perhaps the answer is beyond the current resources of mathematics. But there is another, stronger twin prime conjecture that states that the number of twin primes less than x grows close to another fully calculable number that grows without limit and depends on x. Clearly, this strong twin prime conjecture implies the usual twin prime conjecture. The first few pairs of twin primes are (3,5), (5,7), (11,13), (17,19), (29,31), (41,43), (59,61), (71,73), (101,103). Today, the largest known twin primes have more than 24,000 digits. It is interesting to note that in 1995 T. R. Nicely used the twin primes 824,633,702,441 and 824,633,702,443 to discover a flaw in the Intel Pentium microprocessor.

As with the Goldbach Conjecture, after the last edition of *Number* was published, a flood of theorems converged toward the twin prime conjecture. Since 1919, we knew that there are infinitely many numbers k such that both k and $k + 2$ are products of at most nine primes. Just after the last edition of *Number*, it was discovered that k and $k + 2$ are products of at most three primes.

Computer programmers building tests, giving machines heated workouts, are hitting many of these conjectures, optimistically searching for more twin primes or zeros of the zeta function. Why do they bother? No matter how many twin primes or zeros they find, they could never prove the conjectures that way. They are not trying to prove anything, but rather trying to display what theorists believe exists. Each new find contributes to confidence in the conjecture. Pessimists would hope to find a zero of the zeta function off the magic line to give a counterexample. That's possible. But if the first 1,022 zeros follow Riemann's prediction, how likely would it be that the next will not? And then we must ask this question: Riemann checked only the first three zeros, so how could he have possibly known that they would all lie on the line with real part equal to 1/2? Answer: He knew something about the character, purpose, and destination of the whole beast, not just what it is when it stops to pick up another zero.

This limited selection is a sampling of some of the countless jewels of mathematics that were advanced in the past 50 years. The choices here are limited to the subjects treated in *Number* and hence more connected to the field of number theory. However, readers of *Number* should be aware that although few of the prize problems mentioned in *Number* have been solved, the past 50 years of attempts at solving problems like them have given us a higher—much higher—comprehension of the things we do when we do mathematics. We now see it all coming from that one great and stable unifying source—the *thing* that *is* mathematics. This viewpoint was unavailable to Dantzig and other mathematicians working in the first half of the twentieth century.

We know also—just as Dantzig did back in 1954—that great theorems of mathematics tidily unveil themselves in one branch to cast teasing silhouettes on delicate curtains separating others. Perhaps some curtains will gently separate in the breeze of the next 50 years.

—Joseph Mazur

Notes

1   *the bird can distinguish two from three*   Even animals possess some crude "knowledge" that permits them to distinguish number. In the 1930s, the experimental ethnologist Otto Koehler and others at the University of Freiburg hypothesized that to be able to count, one must be able to simultaneously compare collections of objects and to remember numbers of objects. In one of his amazing experiments designed to test whether animals can deal with numbers, Koehler was able to train a raven to distinguish numbers of spots, from two to six. On learning these numbers, five boxes were set out labeled with 1, 2, 1, 0, and 1 spots, respectively. The raven first opened the first three boxes, consumed four pieces of food, and then left. The raven, "thinking" that he must have made a mistake, returned to recount the spots by bowing its head once before the first box, twice before the second, and once again before the third. He then passed over the fourth box to the fifth, opened it and consumed the fifth piece of food.

Koehler hypothesized that the bird had made what he called "inner marks," somewhat like what humans do when they count on their fingers. He believed that there must be some marking mechanism that enabled the raven to recognize and record number. We know that

birds can count eggs in their nest and lay only as many eggs
as they can support with food. A bird will eat an egg if
there is one too many and lay another if it can support
another.

4   *no such faculty has been found among mammals* Lions in
the wild can compare sizes of prides; they will attack only
if they outnumber the intruding pride. They can distinguish size by numbers of distinct roars. Thus they have a
sense of size comparison. Koehler believed that all animals,
including humans, have some "marking" scheme to keep
track of number sense.

4   *so limited in scope as to be ignored* A 1992 article in the
journal *Nature* (Karen Wynn. "Addition and Subtraction
by Human Infants," *Nature*, [1992] 358: 749–50) reported
on experiments that show that five-month-old infants
possess some crude number sense: A group of five-month-
old infants was shown a puppet (in the image of Mickey
Mouse) being placed on a stage. A screen was dropped to
hide the puppet from view. The infants were then shown a
second puppet being placed behind the screen. The screen
was then lifted. If two puppets were visible, the infants
showed no surprise , as measured by the amount of time
the infants stared at the scene; but if only one puppet was
visible, the infants showed surprise. (Psychologists measure infant surprise by measuring the length of time a baby
stares at an object. If the baby stares at an object for a short
time and then looks away, we may infer that the baby has
seen that object before and has become bored. A prolonged
stare suggests that the baby has not had the experience of
seeing what it is seeing and is therefore putting this new
experience into its proper category of associations.)
Experiments similar to Karen Wynn's have been carried

out with rhesus monkeys in the wild with the same results. (See Hauser, M., MacNeilage, P., and Ware, M. [1996]. Numerical representations in primates. Proceedings of the National Academy of Sciences, USA, 93, 1514–17.)

Another group of infants was shown two puppets before the screen rose to hide the puppets from view. The screen was then lifted. If one puppet was visible, the infants showed no surprise; but if two puppets were visible, the infants showed surprise. This suggests that five-month-old infants could subtract one from two.

Could it be that the infants in Karen Wynn's experiment were simply forming mental images to notice whether there are missing puppets, just as a bird could notice an egg missing from her nest? Experiments by Etienne Koechlin were designed to test whether the infants were forming mental models of the objects (Stanislas Dehaene, *The Number Sense*, pp: 55–56, Oxford University Press). Koechlin's experiment was similar to Wynn's, except that the stage rotated slowly so that the positions of the puppets could not be predicted and therefore not fixed as a mental image. Koechlin found that the infants were still surprised when the screen lowered and the incorrect number of puppets was shown, thus demonstrating that the infants were not using precise mental images of the configuration behind the screen.

From another experiment it was determined that, amazingly, the infant's "computations" are independent of object identity (Simon, T. J., Hespos, S. J., and Rochat, P. [1995]. Do infants understand simple arithmetic? A replication of Wynn [1992]. *Cognitive Development*, 10, 253–269). Infants were still surprised when the incorrect number of objects appeared, but not surprised to find the puppets replaced by balls, thus demonstrating the infant's abstract number cognition.

9 *turn down these fingers in succession* The Yupno, an Aboriginal tribe living in the remote highlands of New Guinea, count to 33 using an elaborate system that counts each finger in a given order, then notes body parts, alternating from one side to the other, including toes, ears, eyes, nose, nostrils, nipples, belly button and genitals (Wassmann, J., and Dasen, P. R. [1994]. Yupno number system and counting. *Journal of Cross-Cultural Psychology*, 25[1], 78-94).

10 *full instructions in the method* The only complete record of ancient finger counting in existence is the codex *De computo vel loquela digitorum*, "On Calculating and Speaking with Fingers," written by Venerable Bede, an eighth-century Benedictine monk renowned among medieval scholars for, among other things, his calculation of the varying date of Easter Sunday, which was designed to never fall on the same day as the Jewish Passover. Because all other Church holidays are determined by Easter, Bede's calculations were considered significant. Bede illustrates how one can indicate numbers from 1 to 1 million by simply extending and bending fingers. (For further details, see Karl Menninger, *Number Words and Number Symbols: A Cultural History of Numbers*, Dover, [1992], New York, pp 201–220.)

47 *five more have been added to this list* Today 39 perfect numbers are known, the largest being $2^{13,466,916}(2^{13,466,917} - 1)$. It contains more than 4 million digits. Of course, it may not be the thirty-ninth perfect number.

50   *twin primes*   In 2000, the largest known twin primes were
      discovered. They are $665551035 \times 2^{80025} \pm 1$. They have
      24,098 digits. In November 1995, twin primes were used
      to reveal a flaw in the Intel Pentium microprocessor, which
      should have been accurate to 19 decimal places but were
      incorrect after the tenth. (Source: Eric W. Weisstein. "Twin
      Primes." From MathWorld—A Wolfram Web Resource.
      http://mathworld.wolfram.com/TwinPrimes.html.)

52   *must be a multiple of five*   To see that $n(n^2 + 1)(n^2 - 1)$ is
      always a multiple of 5, factor the last term. The product
      may then be rearranged as $P(n) = (n - 1)(n)(n + 1)(n^2 + 1)$. Notice the following: $P(n)$ is divisible by 5, if any one of
      the first three factors is divisible by 5. If none of those first
      three is divisible by 5, then n must leave a remainder of
      either 2 or 3 after being divided by 5. If n leaves a remain-
      der of 2, then n^2 leaves a remainder of 4. If n leaves a
      remainder of 3, then n^2 leaves a remainder of 9. In each of
      these last two cases, adding 1 to the remainder gives a
      remainder of 5, so $n^2 + 1$ is divisible by 5.

53   *still challenging the ingenuity of mathematicians*   The
      postulate of Goldbach, more commonly known as the
      Goldbach Conjecture, continues to be one of the world's
      great unsolved problems. It is one of the oldest unsolved
      problems in number theory. It was conjectured in a 1742
      letter from the Prussian mathematician Christian
      Goldbach to Leonhard Euler. It has been verified for all
      numbers less than 6×10^{16} by the Portuguese mathemati-
      cian Toma[s]s Oliveira e Silva, at the University of Aveiro.

55 *the general proposition has not been proved*   Andrew Wiles proved the general proposition in 1994.

65 *which corresponded to our 100,000*   Dantzig means 100,000,000. The name *octade* refers to the exponent in 10^8.

68 *the model for all exact sciences*   Geometry has been a model of deductive reasoning since Euclid's time, but that model is not limited to geometry. Deductive reasoning in number theory had a rich reputation for deductive proof long before the nineteenth century, when axioms for arithmetic were established.

70 *What is the principle involved?*   Dantzig means to say that occasionally deductive methods are not enough and that there is another powerful principle that mathematicians use as a tool to prove their theorems. This point is clarified later on in the chapter, when Dantzig applies the principle of mathematical induction to show the associative law of arithmetic.

72 *an unfinished match of two gamblers*   The problem was known as the *problem of points*. It asked for the number of points that should be awarded to each of two players in a game of dice if the game is left unfinished. Originally, the problem was stated by Girolamo Cardano in an unpublished Latin manuscript, filled with important contributions to calculating probabilities connected with gambling, titled *Liber de Ludo Aleae* (Book of Dice Games). Cardano was a Milanese physician, mathematician, and gambler, better known for his 1545 published book, *Ars Magna* (The Great Art), an account of everything known about the theory of algebraic equations up to the time.

80   *the papyrus Rhind*   This was part of a scroll written about 1700 B.C. and discovered in 1858 by Henry Rhind, a Scottish antiquary.

80   *Ahmes*   The scribe is A'h-mosé who lived sometime between the fifteenth and seventeenth centuries B.C. It is believed that A'h-mosé copied the work from an eighteenth-century B.C. work.

82   *syncopation of the first syllable of the Greek word* arithmos Dantzig must mean the *last* syllable.

90   *liberated algebra from the slavery of the word*   For a wonderful, deeper, scholarly account of Dantzig's point, see Jacob Klein, *Greek Mathematical Thought and the Origin of Algebra*, MIT Press, (1968), Cambridge, MA, pp. 150–185.

94   *no rational number which satisfies the equation x • 0 = a* This paragraph and the next give remarkable reasons for why we cannot divide by zero. However, the one thing that should be said is that zero is the unique number that is the product of any number with zero. Therefore, if *a* is any number but zero, it cannot be equal to x • 0.

97   *in the form of a couple*   The list given here is not meant to give the impression that the lower number is always 1. Nor does it mean that the list continues to infinity before listing the lower number as 2. Dantzig simply means a pair of numbers *a* for the numerator and *b* for the denominator. There is an ingenious way to list these numbers by writing a list of these infinite lists in such a way that the *n*th list lists all fractions with denominator *n*. This list of lists can be organized to make a single list of all rational numbers.

106   *Euclid's proof*   The actual proof appears in Euclid, Book X,
      Proposition 9. The actual theorem is ascribed to
      Theaetetus, who proved that the square roots of prime
      numbers from 2 to 17 are also incommensurable with
      unity almost a hundred years before Euclid wrote his
      Elements. In the case of $\sqrt{2}$, the indirect proof shows that
      if $\sqrt{2}$ were commensurable with the side of a square of
      length 1, then there would be a number that is both even
      and odd.

106   *then the diagonal and the side are commensurable*   Two
      lines or the measurements of two distances are *commensu-*
      rable if the ratio of their lengths is a rational number. If
      their ratio is not a rational number, they are called *incom-*
      mensurable.

107   *continued fractions*   For example, $\sqrt{2}$ may be written as
      the so-called *continued fraction*

$$\sqrt{2} = 1 + \cfrac{1}{2 + \cfrac{1}{2 + \cfrac{1}{2 + \cdots}}}$$

108   *I offer this theory for what it is worth*   We may never know
      the method by which the values were attained, but
      Dantzig's theory offers a very simple method and therefore
      was more likely to be the one used.

113   *Diophantus'* Arithmetica   Diophantus wrote in the fourth
      century B.C. His *Arithmetica* was a work on the solution of
      algebraic equations and on the theory of numbers.

113   *rational numbers and quadratic surds*   The cases where B
      was negative were not considered.

114 *by means of radicals only* "By radicals" may seem like an arbitrary requirement, but it turns out that being solvable by radicals is equivalent to saying that the equation is solvable by a finite step-by-step procedure (an algorithm) that can be carried out in a finite amount of time.

118 *built on 8/9 of its diameter* The area of a circle is

$$sR^2 = s\frac{D^2}{4}$$

where R is the radius and D is the diameter. If $(16/9)^2$ is used in place of π, then the area of a circle (expresses as a square) is

$$s\frac{D^2}{4} = \left(\frac{16}{9}\right)^2 \frac{D^2}{4} = \left(\frac{8}{9}\right)^2 D^2$$

118 *these problems* The first equation comes from trying to solve the problem of doubling the volume of a cube by straight edge and compass. The second equation comes from trying to trisect an arbitrary angle by straight edge and compass. This problem requires one to find the quantity $x = \cos(a/3)$. The trigonometric identity $\cos a = 4\cos^3(a/3) - 3\cos(a/3)$ then leads to $4x^3 - 3x - a = 0$.

126 *Zeno of Elea* Zeno of Elea is credited as the inventor of dialectics and should not be confused with the more noted Zeno of Citium, the founder of the Stoic school of philosophy. We know very little about the life of Zeno of Elea. His visit to Athens and a small part of his philosophy is recounted by Antiphon in Plato's dialog *Parmenides*. We get a bit more biography from Diogenes Laertius's *Lives of Eminent Philosophers*, written more than 700 years after

Zeno's death. Zeno's book contained 40 paradoxes on plu-ralism and motion, four of which are preserved in Aristotle's *Physics*.

126 *the Stagyrite*   The Stagyrite is Aristotle, who came from Stagira, an ancient city in Macedonia.

127 *Achilles and the Tortoise*   Aristotle refers to Achilles in a race, but the tortoise seems to have been made up by mod-ern authors to color the story. The main sources for Zeno's arguments are Aristotle's *Physics*, Diogenes Laertius's *Lives of the Philosophers*, Simplicius's *Commentary on the Physics*, and Plato's *Parmenides*. None of these sources talks about a tortoise.

129 *the historical importance of the Arguments*   This historical importance should not be underestimated. Bertrand Russell said, "The problem first raised by the discovery of incommensurables proved, as time went on, to be one of the most severe and at the same time most far-reaching problems that have confronted the human intellect in its endeavor to understand the world," (Bertrand Russell, *Scientific Method in Philosophy*, Open Court, London, ([1914], p. 164).

129 *But the sum of an infinite number of finite intervals is infi-nite*   This is not what Dantzig meant to say. He is putting this in the mouth of Zeno. We know that the infinite sum of consecutive powers of 1/2 is equal to a finite number, namely 1. The statement is true only if the finite intervals are all larger than some finite number.

130 *Dividing time into intervals*   Fifth-century Greeks thought of instants of time, like points on a line, as beads on a string.

147 *sequence*   *Sequence* is a mathematical term meaning a list of objects that follow a definite numeric order. The objects of a sequence are usually numbers.

148 *evanescent*   Here *evanescent* simply means diminishing in numeric value or tending toward zero.

149 *asymptotic*   Two asymptotic sequences have values tending toward each other.

150 *the simplest type of sequence*   *Simplest* in the sense that its terms approach a number that could be explicitly calculated.

151 *the sum of the progression*   This is easily shown by stopping the series at the n + 1st term and calling it S_n.

$$S_n = a + ar + ar^2 + \ldots + ar^n$$

Then computing $S_n - rS_n$, which turns out to be simply $a - ar^{n+1}$.

Solve for S_n in the equation $S_n - rS_n = a - ar^{n+1}$ to find that

$$S_n = \frac{a - ar^{n+1}}{1-r}.$$

Notice that S_n approaches the original geometric series as n gets larger.

152 *sum of an infinite number of terms may be finite*   These are explanations through mathematical models that seem to work. But the paradoxes remain. If we believe that the models represent the dichotomy and Achilles arguments fairly, then Zeno's arguments are reduced to riddle stature.

But there is always the question of how the continuity of
motion can be represented by those strange pauses of halving, or by those peculiar brakes in the race between
Achilles and the tortoise.

155   *a new type of mathematical being*   This may seem like a
strangely circuitous way of representing what we know as
a real number, but the reader will see that these so-called
self-asymptotic convergent sequences lend themselves to a
solid definition of what it means to be a real number having all the properties of arithmetic and continuity that one
would expect of the set of real numbers. Notice that such a
definition is built on the understanding that the rational
numbers are already defined. It uses the definition of
rational number, what it means to be asymptotic and convergent, and what it means to be a sequence.

157   *which form the convergent sequence 1, 1.4. 1.41. 1.414.*
1.4142, 1,41421   The reader may question the system that
continues this sequence. For example, what is the seventh,
eighth, etc. term in the sequence and how is it constructed?
The answer is that there are several algorithms for con-
structing what we expect will be a sequence of numbers
converging to v $\sqrt{2}$. The reader will come across one later
in the chapter.

161   *already known to the Greeks*   The infinite process that
leads to the same result as continued fractions was known
to the Greeks. See David Fowler's *The Mathematics of
Plato's Academy*, Second Edition, Oxford University Press
(1999), and Wilber Knorr's *The Evolution of the Euclidean
Elements: A Study of the Theory of Incommensurable
Magnitudes and Its Significance for Early Greek Geometry,*
Kluwer Academic Publishers Group (1980).

161   *we obtain the fraction*   Dantzig means this to be written as a fraction with each term being in the denominator of the denominator of the preceding term. It should be illustrated as follows:

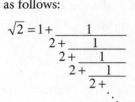

$$\sqrt{2} = 1 + \cfrac{1}{2 + \cfrac{1}{2 + \cfrac{1}{2 + \cfrac{1}{2 + \ddots}}}}$$

164   *as a special case of continued*   Should be "as a special case of continued fractions."

165   *slowly but surely diverges*   Note that

$$\frac{1}{3} + \frac{1}{4} < \left(\frac{1}{4} + \frac{1}{4} \right)$$

$$\frac{1}{5} + \frac{1}{6} + \frac{1}{7} + \frac{1}{8} < \left(\frac{1}{8} + \frac{1}{8} + \frac{1}{8} + \frac{1}{8} \right)$$

etc.

to see that the harmonic series

$$1 + \frac{1}{2} + \left(\frac{1}{3} + \frac{1}{4} \right) + \left(\frac{1}{5} + \frac{1}{6} + \frac{1}{7} + \frac{1}{8} \right) \cdots$$

is greater than

$$\frac{1}{2} + \left(\frac{1}{4} + \frac{1}{4} \right) + \left(\frac{1}{8} + \frac{1}{8} + \frac{1}{8} + \frac{1}{8} \right) + \cdots,$$

which is equal to

$$\frac{1}{2} + \frac{1}{2} + \frac{1}{2} + \cdots.$$

This last series grows indefinitely and hence diverges.

166 *rearrange the terms at will* The simple case referred to on page 166 shows that associativity and commutativity are not valid for series with both positive and negative terms. Pick any non-zero number a and form the sequence of $a - a + a - a + \ldots$. Group the terms as $(a - a) + (a - a) + \ldots$ and the result is 0. But if the terms are grouped as $a + (-a + a) + (-a + a) + \ldots$, the result is a.

167 *the natural logarithm of 2 as a limit* Dantzig seems to have made a mistake here.

$y = 1 + 1/2 + 1/4 + 1/6 + \ldots = 1 + 1/2(1 + 1/2 + 1/3 + 1/4 + 1/5 + \ldots)$. And from this we carelessly draw

$y = 1 + 1/2(x + y)$ or $y - x = 2$. Then we should "fallaciously" conclude that the alternating harmonic series converges toward 2, not 0. His point still holds because we would still arrive at a fallacious conclusion.

174 *does not constitute a continuum* For example, we have seen that the limiting value of the sequence

$(3/2)^2, (4/3)^2, (5/4)^2, (6/5)^2 \ldots$ is the transcendental number e, which is not rational.

182 *analogous to the infinite sequence of Cantor* It is often the case in mathematics that two seemingly different theories give the same result. If one seems to avoid a certain notion (say infinity) and the other depends on it, then they both really need the notion, although one is happy to disguise it and pretend it is not being used.

190 *lead to the impossible expression* To see this, recall a bit of algebra and the quadratic formula.

Denote the two parts as x and y. Then $x + y = 10$ and $xy = 40$.

Therefore $x = 10 - y$. Substituting this expression for x into the equation $xy = 40$ gives $(10 - y)y = 40$, which simplifies to $y^2 - 10y + 40 = 0$. Now use the quadratic formula to find $y = 5 \pm \sqrt{15}$.

191 *leads to a purely illusory result* Cardono found a formula for solving equations of the form $x^3 + ax + b = 0$. His formula gave

$$x = \sqrt[3]{\frac{-a}{2} + \sqrt{\frac{a^2}{4} + \frac{b^3}{27}}} + \sqrt[3]{\frac{-a}{2} - \sqrt{\frac{a^2}{4} + \frac{b^3}{27}}}.$$

Applying this formula to the equation $x^3 + 15x + 4 = 0$ yields the solution

$$x = \sqrt[3]{2 + \sqrt{-121}} + \sqrt[3]{2 - \sqrt{-121}}.$$

192 *With this notation the solution of Bombelli's equation is* To see this, first notice that $\sqrt{-121} = 11i$ Then use the fact that $i^2 = -1$ to show that

$$(-2 \pm i)^3 = 2 \pm 11\, i.$$

This may be shown by simply multiplying $(2 \pm i)$ by itself three times and simplifying with the rule that $i^2 = -1$.

Use this last fact to show that

$$\sqrt[3]{2 + 11i} + \sqrt[3]{2 - 11i} = (2 + i) + (2 - i) = 4.$$

193 *striking identity of Euler* Of course, one must question what raising a number to a non-integer power (or worse, to a complex power) could possibly mean. It is clear what x^n means when n is positive whole number. The meaning is extended and generalized by keeping the natural rules of powers intact. For example, $x^{n+m} = x^n x^{\,m}$. This rule provides a framework for extending the definition of what it means to raise a number to an arbitrary real number and

then, using both the identity of de Moivre (a conversion of complex number powers to sines and cosines) and an interesting way of viewing complex numbers on a plane (the so-called complex plane), it provides a way of extending the definition to what it means to raise a real number to a complex power.

200 *their profound and far reaching thoughts*   The foundations of calculus.

202 *the mean proportional*   The mean proportional of two quantities is the square root of the product. In this case the two quantities are L and y. So the language of Latin terms (which are no longer used) translates into saying that the semichord x is equal to the square root of the *latus rectum* L and the height y.

210 *the vectors* OA *and* OB   For readers not familiar with vectors, think of a vector as a pair of numbers that could also be thought of geometrically as a line starting at O and ending at A. In this case, the point O is the point at the origin with coordinates $(0,0)$. If A represents $a + ib$, then, in ordinary Cartesian geometry notation its address is (a,b). So the vector is the line starting at $(0,0)$ and ending at (a,b). There are several advantages to reinterpreting the point A as a vector. One is that two vectors can be added. The vector OA can be added to the vector OB by simply adding their components. If OA ends at the point (a,b) and OB ends at (c,d), then the sum $OA + OB$ ends at $(a + c, b + d)$. This is the parallelogram rule of addition of vectors, so-called because the new vector turns out to be the diagonal of a parallelogram with sides OA and OB.

220 *The modern theory of aggregates*   Interpret *aggregates* to mean *collection*.

222 *play in the arithmetic of the finite*   This can be taken as the definition of being infinite. A set is said to be infinite if its members can be paired one-on-one with members of a proper subset of itself. *Proper* here means that there is at least one member of the subset not included in the larger set. For example, the set of even positive integers is a proper subset of the set of all positive integers because 3 is a member of the larger set not included in the set of all even integers. The set of positive numbers is infinite, because every integer can be paired with the integer twice its value, so every integer has a buddy that is even.

224 *denumerable*   The term *countable* is often used in place of *denumerable* because, in the case of a finite collection of things, counting is the process of setting up a one-to-one correspondence with a finite subset of the positive whole numbers.

225 *See figure, page 225*   There is an alternative way of seeing this. Build the array differently. Display on the infinite line all the positive whole numbers. On the next line display the same infinite line of whole numbers, but this time make them fractions with each one divided by 2. On the third line, display the same infinite line of whole numbers replacing all the denominators by 3, etc. You should have the array illustrated below. Every fraction *p/q* may be placed on a two dimensional array, addressed by its vertical and horizontal position—*q* places to the right and *p* places down. Use the serpentine arrows to count the rational numbers, skipping any that have already been counted.

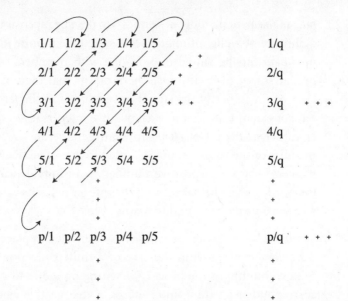

228   *there is only one equation of height 1*   For small heights,
      there are few choices for the coefficients. For height equal
      to 1, the polynomial cannot be of degree greater than 1. So
      if one looks at the possibilities for polynomials of degree 1,
      there is only one choice: $x = 0$. For height equal to 2, the
      only possibilities are $x + 1 = 0$, $x - 1 = 0$ and $x^2 = 0$. For
      height equal to 3, start with equations of degree 1 and
      work your way toward the higher degrees. Notice that the
      scheme stops at the one and only equation of degree 3,
      namely $x^3 = 0$. There are five: $x + 2 = 0$, $x - 2 = 0$, $2x + 1 =$
      0, $2x - 1 = 0$, $3x = 0$.

229   *the diagonal procedure*   When you try to enumerate the
      collection of all real numbers you will find that there is a
      problem. Real numbers between 0 and 1 are numbers that
      can be represented by a (possibly infinite) string of deci-
      mals. For example 0.4673904739828983493... is one such

number where the three dots indicate the digits go on forever. Just try to list real numbers from 0 to 1. One possible
list may look like this:

1   0.46739 $\cdots$
2   0.38654 $\cdots$
3   0.03936 $\cdots$
4   0.84534 $\cdots$
5   0.67657 $\cdots$
     $\vdots$

No matter how you arrange these real numbers there will
always be infinitely many that are not on your list. Here is
just one: Take the number you get by reading down the
infinite diagonal of the infinite array (diagonal encir-
clement below). The number you get is 0.48937....

1   0. 4 6 7 3 9 $\cdots$
2   0. 3 8 6 5 4 $\cdots$
3   0. 0 3 9 3 6 $\cdots$
4   0. 8 4 5 3 4 $\cdots$
5   0. 6 7 6 5 7 $\cdots$

n-th digit of n-th
number on the list

n   d

Now construct a new number by changing this diagonal
number as follows: If a digit is not 9, add 1 to it. Change
any 9 into a 0. In this example, the first digit becomes 5, the
second becomes 9, the third becomes 0, and so on, so the
newly constructed number is 0.59048.... This number is
not on the list. If it were, it would have to be somewhere on
the list, say in the nth position, and we would come to the
most bizarre situation of having a digit d, the nth digit of
the nth number on the list (the one in the circle) that is

both d and $d + 1$ at the same time. The only recourse is to admit that the number we constructed (0.59048...) is not on the list. This shows that the set of real numbers is *larger* than the set of integers and hence *larger* than the set of rational numbers. The digits along the diagonal go on forever, so infinitely many numbers could have been constructed in this way giving infinitely many not on the list. (We could have added 2 or 3 or any number between 1 and 9 to the digits of the diagonal number and achieved the same result.) This shows that the cardinality of the set of real numbers is greater than the cardinality of the set of rational numbers.

230   *the longer line contains no more points than the shorter*   To see this, draw two lines AB and CD; put the shorter one AB above the longer. Connect the respective endpoints to form two lines AC and BD and extend them upward until they meet at a point P. Then pick any point Q on AB, draw a line from P to Q, and extend PQ to intersect CD at R. In this way you have made a one-to-one correspondence between the points on AB and the points on CD. (See illustration below.)

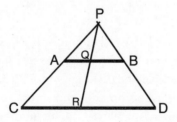

231   *decimal fractions*   Decimal fraction here means simply decimal expansion of the real number between 0 and 1. For example, the real number $1/\pi = 0.31831\ldots$.

233   *have so far not been crowned with success*   The question of whether such a set exists refers to the Continuum Hypothesis. Dantzig does not mention the Continuum Hypothesis by name, although he talks at length about the continuum. Before 1964, there was no answer to the question of whether there is a set whose cardinal is greater than that of the integers and less than that of the real numbers. The Continuum Hypothesis claimed that there was no such set. Paul Cohn, a young mathematician at Stanford, settled the question in 1964, showing that the Continuum Hypothesis is neither true nor false. Put another way, its truth is not decidable because it depends on the system of axioms chosen for set theory.

233   *totality of all correspondences*   These correspondences may be considered as a set of objects.

237   *entirely dependent on the concept of natural number*   One big difference between the two camps should be noted here. The formalists agree that a mathematical object exists if that existence does not lead to a contradiction, but the intuitionist only accepts mathematical objects that can be described or constructed in a finite number of steps.

Further Readings

Except for very few out-of-print suggestions, the following were picked because of their clarity, accessibility, and availability. Most are available through bookstores. Others are available through good libraries. With so many good, relevant books available, the task of choosing is difficult, but here is a limited list.

As an overall sourcebook consider:

Courant, Richard, and Herbert Robbins. What Is Mathematics? An Elementary Approach to Ideas and Methods. Revised by Ian Stewart. Oxford: Oxford University Press, 1996.

Like *Number*, this book was originally written before World War II; it remains a classic introduction to what mathematics does. It provides a good one-stop source for many of the technical questions encountered in *Number*. It entices curious readers to pursue deeper understanding of the wonderful collection of interesting topics it explores from many branches of mathematics.

Chapter 1: Fingerprints

Butterworth, Brian. *What Counts: How Every Brain Is Hardwired for Math*. New York: Free Press, 1999.

This is an extraordinary account of how we and other animals think of numbers and mathematics.

Dehaene, Stanislas. *The Number Sense: How the Mind Creates Mathematics*. New York: Oxford University Press, 1997.

A readable, interesting account of how humans (and animals) think about mathematics. In particular, Chapter 2 is about the number sense of newborns, and Chapter 4 is about how humans conceptualize numbers.

Menninger, Karl. *Number Words and Number Symbols: A Cultural History of Numbers*. Translated by Paul Broneer. New York: Dover, 1992.

This book is extremely comprehensive and filled with the details of the cultural history of numbers. Read any section and become absorbed in wonderful facts about the evolution of number writing, symbols, and cultural notions of counting. The book is filled with photographs and drawings of ancient counting, calculating, and measuring artifacts.

Neugebauer, Otto. *The Exact Sciences in Antiquity*. New York: Dover, 1969.

Ever since these lectures were delivered at Cornell University in 1949, they have stood as the standard history of Egyptian and Babylonian mathematics. This book is very readable and at the same time referred to by scholars of the history of science.

Chapter 2: The Empty Column

Cajori, Florian. *A History of Mathematics*. New York: Chelsea, 1999.

This is a compact history of mathematics from antiquity to the early part of the twentieth century. This book is an excellent companion to *Number*, written in a comparably beautiful style.

Kaplan, Robert. *The Nothing That Is: A Natural History of Zero*. New York: Oxford University Press, 1999.

This is a truly witty and entertaining exposition examining the evolution of the concept of zero. Kaplan begins by saying, "If you look at zero, you see nothing; but look through it and you will see the world." Read this book to see the world differently.

Seife, Charles. *Zero: The Biography of a Dangerous Idea*. New York: Viking, 2000.

This is another humorous book on zero. Aside from its more matter-of-fact style, the one big difference between this and Kaplan's book is that it spends much more time on physical consequences of zero and infinity.

Chapter 3: Number Lore

Aczel, Amir. *Fermat's Last Theorem: Unlocking the Secret of an Ancient Mathematical Problem*. New York: Delta, 1997.

This wonderful adventure tale gives readers a remarkable sense of the background to one of the world's most famous problems. Aczel's art is in explaining that background through high school level mathematics without trivializing the story.

Allman, George. *Greek Geometry from Thales to Euclid.*
New York: Arno Press, 1976.

This nineteenth-century work complements Proclus's
book, which, in turn, summarized a lost history of
geometry written by Eudemus in the third century B.C.
It contains a comprehensive, well-written commentary on
fifth-century B.C. mathematics surrounding Pythagorean
mathematics.

Berlinghoff, William, and Fernando Gouvèa. *Math
Through the Ages: A Gentle History for Teachers and Others.*
Farmington, MA: Oxton House, 2002.

This is a very well-organized and friendly book. It is broken into two parts: a general overview, and "sketches" providing more detail. The beauty of this book is in its exposition: The overview does not interrupt the flow with details
that are continuously linked to the sketches. One is
reminded of Internet reading with highlighted words sig-
naling links to greater detail.

Ogilvy, Stanley, and John Anderson. *Excursions in Number
Theory.* New York: Dover, 1966.

This is a wonderful primer in number theory creatively
filled with interesting ideas and problems. The authors
present us with stimulating material of substance while
carefully balancing technical treatment with accessibility.

Ore, Oystein. *Number Theory and Its History.* New York:
Dover, 1968.

Written with little technical language, this book outlines
the history of number theory from its earliest beginnings
to the point where modern machinery takes over. It is

written in a clear and concise style, touching many alluring points of the subject.

Singh, Simon. *Fermat's, Enigm : The Epic Quest to Solve the World's Greatest Mathematical Problem.* New York: Anchor, 1998.

This is the most popular account of the recent proof of Fermat's Last Theorem. Though thorough understanding of the proof requires extraordinarily advanced knowledge of mathematics, this book presents the lively story of the of the remarkable problem and, using only high school level mathematics, gives readers an impressive sense of the adventure.

Whitehead, Alfred North. *An Introduction to Mathematics.* New York: Henry Holt, 1939.

This small book is a collection of important basic ideas necessary to learning mathematics. It is a bit out of date but is still very readable.

Chapter 4: The Last Number

Goodstein, R. L. *Essays in the Philosophy of Mathematics.* Leicester, UK: Leicester University Press, 1965.

This is a compilation of very readable essays reprinted from several respected journals by Reuben Goodstein, a prolific writer known for his clear, expository style. For readers who want to quickly understand the notions of proof and the axiomatic method without much work, the essays in this book are excellent.

Chapter 5: Symbols

Cajori, Florian. *A History of Mathematical Notations.* New York: Dover, 1993.

This book was originally published in 1929. It gives a comprehensive account of how ancient and obsolete notation evolved into more modern notation.

Hobgen, Lancelot. *Mathematics for the Million.* New York: Norton, 1946.

An eclectic collection of mathematical topics. Chapter 7, "How Algebra Began," is a beautifully written book report on Dantzig's *Number*, starting with matchstick number symbols used by the Chinese in the first century B.C. and developing, through practical problems, into the algebraic shorthand of sixteenth-century algebra.

Chapter 6: The Unutterable

Beckmann, Petr. *A History of Pi.* New York: St. Martins Griffin, 1996.

Forget the political innuendos against the Soviet Union and slightly inaccurate statements to find an interesting, funny, and readable book.

Niven, Ivan Morton. *Numbers: Rational and Irrational.* Washington, DC: Mathematical Association of America, 1961.

This is an excellent introduction to the irrational numbers, starting from the natural numbers. The early chapters are elementary and the later for more advanced and ambitious readers.

Russell, Bertrand. *The Principles of Mathematics*. London: George Allen & Unwin, 1956.

This book was written at the turn of the twentieth century, but it is still one of the best, most clear accounts of the philosophy of mathematics that can be found. Chapter 33 of Russell's book corresponds closely with Chapter 6 of Dantzig's book. This 500-page book is a comprehensive account many of the topics discussed in *Number*.

Chapter 7: This Flowing World

Russell, Bertrand. *Our Knowledge of the External World*. Chicago: Open Court, 1914.

If you can get your hands on this book, it is well worth going through Chapters 3, 4, and 5. Russell writes in his inimitable smooth style and gives one of the clearest understandings of Zeno's arguments available.

Chapter 8: The Act of Becoming

Russell, Bertrand. *The Principles of Mathematics*. London: George Allen & Unwin, 1956.

This is the same book suggested for Chapter 6. Chapter 32 is the chapter that corresponds to Dantzig's Chapter 8. There are a few dense mathematical moments in this chapter, but Russell has a clear writing style that carries the reader through much of the harder ideas.

Chapter 9: Filling the Gaps

Dedekind, Richard. *Essays on the Theory of Numbers*. New York: Dover, 1963.

In this thin book, one learns firsthand Dedekind's theory of irrational numbers. Here is where the precise and rigorous definition of irrational occurs. The book also explains Dedekind's view on transfinite numbers and continuity. It was originally written at the end of the nineteenth century in awkwardly concise wording. So it takes a bit of work to understand, but it is well worth the effort.

Chapter 10: The Domain of Number

Mazur, Barry. *Imagining Numbers (Particularly the Square Root of Minus Fifteen)*. New York: Farrar Straus Giroux, 2003.

An easily readable, enlightening account of imagination in poetry and mathematics. Chapter 2, "Square Roots and the Imagination," and Chapter 3, "Looking At Numbers," have particular relevance to this chapter.

Reichmann, W. J. *The Spell of Mathematics*. London: Penguin, 1972.

As the title suggests, one does fall under a spell reading this book. Reichmann associates topics and shows their surprising links. Read the first seven chapters and then jump to Chapter 15, "What's It All About?"

Sawyer, W. W. *Mathematicians Delight*. London: Penguin, 1976.

If you want a gripping introduction to an eclectic assortment of topics in mathematics, try this one. Sawyer gives plenty of examples to help the reader bridge the gap from theory. This covers many of the topics alluded to in *Number*.

Stewart, Ian. *Concepts of Modern Mathematics*. New York: Dover, 1995.

This is precisely about what the title says it is about. If you have ever read other books by this author, you will know that the reading will be clear, concise, accurate, current, and lucid.

Whitehead, Alfred North. *An Introduction to Mathematics*. New York: Henry Holt, 1939.

This small book is a collection of important basic ideas necessary to learning mathematics. It is a bit out of date but is still very readable.

Chapter 11: The Anatomy of the Infinite

Aczel, Amir. *The Mystery of the Aleph: Mathematics, the Kabbalah, and the Search for Infinity*. New York: Washington Square Press, 2000.

A gripping book tracing the development of infinity through the life of Cantor.

Bolzano, Bernard. *Paradoxes of the Infinite*. Translated by Dr. Fr. Prihonsky. London: Routledge and Kegan Paul, 1950.

I include this reference for the historical introduction by Donald Steele, which amounts to almost half the book. The language is archaic, but the historical details create a strong picture of a period when infinity was actively being studied. Bolzano's book has hundreds of highly interesting paradoxes of the infinite that are quite accessible to anyone without a math background who wants to wade through the archaic language.

Gamow, George. *One Two Three... Infinity: Facts and Speculations of Science.* New York: Viking, 1961.

Start reading, and you will find that it is hard to stop. Although this book was written in the middle of the last century, it still has a magnificent freshness. The book makes surprising connections between eclectic branches of mathematics and science. There is plenty here to introduce you to infinity.

Kaplan, Robert, and Ellen Kaplan. *The Art of the Infinite: The Pleasures of Mathematics.* New York: Oxford University Press, 2003.

Like Kaplan's other book, *The Nothing That Is,* this witty book is a page-turner. Kaplan's poetic writing style brings a unique pleasure to novices reading about mathematics. It is an excellent elementary, compressive treatment of the infinite.

Maor, Eli. *To Infinity and Beyond.* Princeton, NJ: Princeton University Press, 1991.

This is a beautifully written primer on infinity. Maor carefully escorts his readers through delicate questions of convergence of infinite series; limits; paradoxes involving infinity; infinite tiling questions posed by M. C. Escher's drawings; and notions of infinity applied to ancient and modern cosmologies.

Lavine, Shaugham. *Understanding the Infinite.* Cambridge: Harvard University Press, 1998.

In this book, Lavine gives original ideas surrounding the philosophy and history of infinity. Parts are accessible to the general reader, but much of this book is addressed to mathematically sophisticated audience.

Rucker, Rudy. *Infinity and the Mind*. Princeton, NJ: Princeton University Press, 1995.

This is a beautifully written book involving many aspects of infinity, clarifying Cantor's arguments and exploring infinity in all its forms from different points of view. This book includes a very clear exposition of Gödel's incompleteness theorems.

Russell, Bertrand. *Introduction to Mathematical Philosophy*. London: George Allen and Unwin, 1919.

This is out of print but may be easily found in most good libraries. It is an amazingly clear exposition of the foundations of natural numbers. In just three short chapters, Russell, in his inimitable style, gets to infinity and induction.

Chapter 12: The Two Realities

Few books deal with alternatives to Dantzig's view on the answer to what is mathematics. The following come close, but it would be nice to have other opinions.

Changeux, Jean-Pierre, and Alain Connes. *Conversations on Mind, Matter, and Mathematics*. Trans. by M. B. DeBevoise. Princeton, NJ: Princeton University Press, 1995.

This marvelous book is a dialog between a biologist and a mathematician on the universality of mathematics and the neurobiology that makes sense of it. There is no other book quite like it that can get readers into the heart of the question of what mathematics is and how we understand it.

Greenberg, Marvin Jay. *Euclidean and Non-Euclidean Geometries: Development and History*. New York: Freeman, 1993.

This is a book about geometry, but Chapter 8 contains a brilliant account and survey of what mathematics is about. Here you will find other relevant references to the question of how mathematics relates to reality. If you want to learn about non-Euclidean geometries, this is the book. It is clearly written and filled with reasonable exercises designed to give the reader an intuitive sense of non-Euclidean worlds.

Lang, Serge. *The Beauty of Doing Mathematics: Three Public Dialogues*. New York: Springer-Verlag, 1985.

This is an unusual collection of dialogues between Serge Lang, a renowned mathematician and an audience of non-mathematicians. Lang points to a woman in the audience and asks, "What does 'mathematics' mean to you?" to begin the dialogue.

Polanyi, Michael. *Personal Knowledge: Towards a Post-Critical Philosophy*. Chicago: University of Chicago Press, 1974.

Polanyi refutes the view that the only valid knowledge is that which can be expressed and tested by strictly impersonal methods.

Wigner, Eugene. "The Unreasonable Effectiveness of Mathematics in the Natural Sciences." *Communications in Pure and Applied Mathematics* 13, No. 1 (February 1960).

This famous essay, referred to in the preface to this edition of *Number*, gives an interesting view of the subjective and objective connections of mathematics with reality.

Index